PRAISE FOR **CLIMATE COVER-UP**

"Absolutely superb—one of the best dissections of the climate information war I have ever seen. This is one terrific piece of work!"

ROSS GELBSPAN
author of *The Heat Is On*

"Through impeccably documented analysis, *Climate Cover-Up* exposes the well-oiled propaganda campaign designed to manufacture dissent and uncertainty about the science of global warming. It is essential reading for anyone who cares about the future of democracy."

ANDREW WEAVER
author of *Keeping Our Cool: Canada in a Warming World*

"A clear and courageous battle cry against those who, for profit's sake, would lead us to environmental and, ultimately, economic ruin."

LESTER R. BROWN
author of *Plan B 3.0: Mobilizing to Save Civilization*

"An important and disturbing book about the lies and corrupt language that government and industry still employ to dismiss the facts on global warming."

ANDREW NIKIFORUK
author of *Tar Sands: Dirty Oil and the Future of a Continent*

"To those of us who have been unknowingly made to turn a blind eye to the terrifying and true facts about global warming, there's no time left for ignorance. Please read this shocking and incredible book, learn how we've been manipulated, get angry, and take action."

NEVE CAMPBELL
actress and producer

"*Climate Cover-Up* reveals how strategic corporate public relations, an unwitting media, and feckless scientists have created a rhetoric-driven public conversation about climate change that defies logic and reason. If you are interested in positive social change on climate issues, this book is a must-read."

FRANKLIN D. GILLIAM JR.
dean, School of Public Affairs, and professor of public policy and political science, University of California, Los Angeles

"Jim Hoggan in this essential book illuminates our folly, even as he points a way forward with hope."

WADE DAVIS
author of *The Serpent and the Rainbow*

"*Climate Cover-Up* clears the way for a new era of honesty and climate progress."

TZEPORAH BERMAN
campaign director and founder, Forest Ethics

CLIMATE COVER-UP

THE CRUSADE TO DENY GLOBAL WARMING

JAMES HOGGAN

with RICHARD LITTLEMORE

CLIMATE COVER-UP

GREYSTONE BOOKS

D&M PUBLISHERS INC.

Vancouver/Toronto/Berkeley

09 10 11 12 13 8 7 6 5 4

Greystone Books
An imprint of D&M Publishers Inc.
2323 Quebec Street, Suite 201
Vancouver BC Canada V5T 4S7
www.greystonebooks.com

Library and Archives Canada Cataloguing in Publication
Hoggan, James, 1946–
Climate cover-up : the crusade to deny global warming/
James Hoggan and Richard Littlemore.
Includes bibliographical references.

ISBN 978-1-55365-485-8

1. Climatic changes. 2. Climatic changes—Government policy.
I. Littlemore, Richard II. Title.
QC903.H63 2009 363.738′74 C2009-903508-1

Editing by Susan Folkins
Copy editing by Eve Rickert
Cover design by Martyn Schmoll
Cover illustration by Martin Barraud/Getty Images
Text design by Naomi MacDougall
Printed and bound in Canada by Friesens
Printed on acid-free paper that is forest friendly (100% post-consumer
recycled paper) and has been processed chlorine free
Distributed in the U.S. by Publishers Group West

We gratefully acknowledge the financial support of the Canada Council for the Arts,
the British Columbia Arts Council, the Province of British Columbia through the
Book Publishing Tax Credit, and the Government of Canada through the Book
Publishing Industry Development Program (BPIDP) for our publishing activities.

In thanks and as a tribute to ROSS GELBSPAN,
whose early scoops and dedicated journalism exposed the
climate change denial campaign and inspired this book.

CONTENTS

ACKNOWLEDGMENTS

It is a rare privilege to have a friend like John Lefebvre, without whose courage, insight, and generosity this book could never have come to pass. We are all deeply indebted to John for his guidance, his constant encouragement, and his ongoing support for the operations of DeSmogBlog.com.

I am also grateful to everyone involved in the DeSmogBlog, especially Richard Littlemore and Kevin Grandia, for their efforts and their research. Many of the details in this book were reported originally on the blog.

The whole community owes a vote of thanks to the scientists and advocates who have worked so hard to catch our attention and build our understanding on the topic of climate change. The Nobel Committee has already offered appropriate praise for former U.S. vice president Al Gore and the scientists of the Intergovernmental Panel on Climate Change. But some of those scientists have taken a particularly public position and have endured unconscionable abuse as a result. Brave and outspoken scientists such as NASA's James Hansen, Penn State's Michael Mann, UC San Diego's Naomi Oreskes, Stanford's Stephen Schneider, and the University of Victoria's Andrew Weaver are among those we know best and respect the most.

There are also scientists and journalists who deserve credit and acknowledgment. Ross Gelbspan, formerly of the *Boston Globe*, was among the first reporters in North America to uncover the extent of the climate cover-up. ABC's Bill Blakemore and the *New York Times*'s Andrew Revkin reported the climate change story accurately when many others were getting it wrong.

In the online world, John Stauber's SourceWatch.org is an encyclopedic font of information, as is Kert Davies's Exxon Secrets.org. Joe Romm has done great work at ClimateProgress .org, and the DeSmogBlog team has long been a fan of Australian online journalist and scientist Tim Lambert, whose Deltoid blog has been a solid source of scoops and thoughtful reporting on the science and politics of climate change.

In the process of assembling the material in this book, the DeSmogBlog received solid support from a host of online sources that also do a great job covering this issue. I'd like to thank and acknowledge Richard Graves of ItsGettingHotIn Here.org; Jesse Jenkins of WattHead.blogspot.com; Pete Altman of Switchboard.nrdc.org; Alex Stefan of WorldChanging.com; Brad Johnson, Faiz Shakir, and Amanda Terkel at ThinkProgress .org; Page van der Linden at DailyKos.com; Drew Curtis at Fark .com; and Andrew Sullivan at AndrewSullivan.TheAtlantic.com.

Given the rigors and distractions that are inevitable in putting together this kind of book, I want to extend a special thanks to some of the people who had to pick up the slack during the long process of research and writing. The whole Hoggan staff has been endlessly supportive, but the greatest thanks must be offered in return for the patience shown by my wife, Enid

Marion, and by Richard Littlemore's whole family, including his wife, Elizabeth, and their three boys, Ted, Avery, and Llewellyn.

Finally, I would like to offer a more specific thanks to Richard Littlemore. I have said before that Richard has a knack for writing down the things I say in the way I wish I had said them. But his contribution to this book went much further. He brought passion, energy, and extensive knowledge of climate change, politics, and journalism. In the earliest days he was the lone pen on the DeSmogBlog, and throughout he has been a tireless researcher and a conscientious reporter. Collaborations of this scope are likely to either ruin friendships or cement them forever. In this case I am delighted to say that I have found and forged a good and lasting friendship.

Jim Hoggan

PREFACE

This is a story of betrayal, a story of selfishness, greed, and irresponsibility on an epic scale. In its darkest chapters, it's a story of deceit, of poisoning public judgment—of an anti-democratic attack on our political structures and a strategic undermining of the journalistic watchdogs who keep our social institutions honest. It is ultimately a story that drove me and those closest to me to outrage and to activism. And although it is not my purpose to make you angry, I hope that you may, through the coming pages, come to understand the sense of indignation and injustice that brought me to write this book.

I didn't go looking for this trouble. I don't think of myself as an activist, and I don't fit the stereotypical description of an environmentalist. I have a decent wardrobe that doesn't include a single hair shirt. I spend too much money on art, fine wine, skis, and high-end bicycle parts, and I am in recovery from my habit of buying luxury cars.

Nor do I bear any grudges against "the establishment"—and particularly not the public relations industry. As the owner of a successful Vancouver public relations firm, I think that PR is a good thing. It connects people and builds understanding, and I generally have a high regard for my professional colleagues. It's true that there have always been bad actors in my business—the tobacco apologists and the partisan political spin doctors—but I have always regarded them as obvious exceptions. In my career, examples of spin-doctoring seemed episodic, not epidemic.

Or that's what I thought before I started looking closely at the climate file. That too began in relative innocence, and only three or four years ago. I was thinking about adding a community service element to the Hoggan & Associates Web site, and somebody suggested a public information section on climate change. I liked the idea immediately. I knew the topic was controversial, and I knew that in a controversy people sometimes oversell their position. I thought it would be useful to introduce an objective viewpoint.

I started doing a lot of reading and was surprised by what I discovered. Where I expected a blistering controversy, I found an overwhelming scientific consensus. Mainstream media had been reporting that doubt lurked in every report, that for every scientist warning of global warming there was another saying it was all bunk. But when I started reading reports from the world's leading science academies, I found that everyone seemed to be speaking with one voice. Every science academy in every major developed country in the world had stated clearly that the world's climate is changing dangerously and humans are to blame. Why, I wondered, were people so confused? Who had started this public debate?

The great U.S. journalist Ross Gelbspan had the answer. In two early books, *The Heat Is On* (1997) and *Boiling Point* (2004), Ross had uncovered the first hard evidence of an organized campaign, largely financed by the coal and oil industries, to make us think that climate science was somehow still controversial, climate change still unproven. I had always known about the potential for public manipulation, but I had never conceived of a campaign so huge, well-funded, and well-organized. Ross is anything but a conventional environmentalist. He's a reporter, skeptical to the bone. And when I flew to Boston to meet him, he told me that when he had started looking into climate change, he actually thought the "science skeptics" had it right. He thought the science was truly stuck in uncertainty. Then Harvard oceanographer Dr. James McCarthy showed Ross how the deniers were twisting the data to mislead people, and he posed what for Ross became an important question: where were these purported skeptics getting their money?

The answer to that question formed the backbone of *The Heat Is On,* and what Ross found struck me as a revelation. Denier scientists were being paid well, not for conducting climate research, but for practicing public relations. As I looked around, I started to notice evidence of the campaign everywhere I looked. To a trained eye the unsavory public relations tactics and techniques and the strategic media manipulation became obvious. The more I thought about it, the more deeply offended I became.

I also found that the same sense of indignation was common among my friends and colleagues. For example, the senior writer at Hoggan & Associates and my collaborator on this book is Richard Littlemore. A veteran newspaper guy, Richard is like Ross Gelbspan, another ink-stained skeptic accustomed to steering a wide berth around anyone who is passionately committed to a cause. But he had been worrying about climate change since 1996, when he took a freelance contract to write a public education package on the topic for the David Suzuki Foundation, Canada's leading environmental organization. Even then, Richard says, reading through the material, "it was clear we were in trouble, and obvious that some people were trying to deny it." In 1998 Richard was selected to be a part of the Canadian government's Kyoto Implementation Process, which he describes now as "a sham," a "vast public relations exercise designed only to waste time—an effort that never had a chance of success."

Richard found himself distraught and disillusioned at the scope and nature of the big lie (in this case, that the Canadian government was serious about reducing national greenhouse gas emissions). It was, he says, built on a foundation of what he came to think of as little insults to democracy, incremental efforts to ensure that government did nothing to disrupt the profitable status quo.

My own gathering horror probably came to a head one day when I started sharing my newfound knowledge with my old friend John Lefebvre, a burly lawyer turned musician who along the way had made his fortune by helping to build an Internet banking empire. John has the kind of money that makes the worries of the world drift into the distance, but he also has a conscience. We were chatting during the summer of 2005 about this corruption of the public conversation when John said, flatly and urgently, "What can we do about it?"

That's how DeSmogBlog was born. We decided to start doing this research in a more organized way and share it with everyone we could find. With a generous stake from John we launched www.DeSmogBlog.com, an unfamiliar but promising Internet platform that we hoped would give us access to a larger audience. Richard started collecting information. He identified people who seemed to be making a living by denying climate change, and he asked a few obvious questions: Were these climate "skeptics" qualified? Were they doing any research in the climate change field? Were they accepting money, directly or indirectly, from the fossil fuel industry? Finding that the most vocal skeptics were *not* qualified, were *not* working in the field, and all too frequently *were* on one or another oily payroll, we started publishing our results online.

From that modest beginning we have built a popular Web site and an active team of researchers and collaborators. We hired Kevin Grandia as a manager early in 2006 and began attracting volunteers such as Emily Murgatroyd—a woman who proved so passionate and determined that we made her part of the team. We engaged brilliant contributors, including the authors Ross Gelbspan, Bill McKibben *(Deep Economy),* and Chris Mooney (*The Republican War on Science* and *Stormworld: Hurricanes, Politics and the Battle over Global Warming*). We found established journalists like Mitch Anderson and hot up-and-comers like Jeremy Jacquot and Nathanael Baker.

More to the point, we assembled the body of research that we share with you here. This is more, however, than a collection of posts or a greatest hits album. We have tried to pull together the whole story, to give you a complete sense of how the public climate change conversation was pushed so badly off the rails.

I suspect that you will find the results offensive, even infuriating. We are at a critical juncture in human history. By mastering technology and by (so far) outperforming every other species on the planet, we humans have achieved global domination. We can remake landscapes, defeat diseases, extend life spans, and expand the scope and scale of human wealth by almost every measure. We can also trash whole countries, pollute streams, rivers, lakes, and perhaps ultimately whole oceans, to a disastrous extent. We can kill one another more quickly than ever in human history, and we can change the world's climate in a way that scientists say is threatening our ability to survive on Earth.

The question, as yet unanswered, is whether we can stop. Can we as a species rescue ourselves from a threat of our own making? To do so will take personal restraint, political courage, and a degree of global cooperation unprecedented in human history. Even more, it will take a clear understanding of the risks—an understanding that we will only achieve if we expose the climate cover-up. That's been our goal, and you may judge our success in your own time. After which, I hope that you will join us in our effort to restore integrity to the public conversation about science, about governance, and about saving the world.

That sounds melodramatic, but I believe two things absolutely. First, I believe that scientists have been telling us the truth when they've said that the world is at risk. And second, even if countering the risk will be difficult, even if the tasks seem overwhelming or the solutions are dismissed by the deniers as impractical, I believe, absolutely, that the world is worth saving.

[*one*]

LEMMINGS AND LIFEGUARDS

Keeping humankind from crashing on the rocks

We are standing at the edge of a cliff. Behind us is a considerable crowd, 6.7 billion people and counting, and below is a beckoning pool. Some people say that you can jump into that pool without risk. They say that humans have been doing so for ages without any problems. But others say that waves have been eating away at the foot of the cliff, causing big rocks to fall into the water. They say that the risk of jumping grows more frightening by the day. Whom do you trust?

That's a tricky question because here, on the climate change cliff, some of the lifeguards are just not that qualified, some have forgotten entirely whose interests they are supposed to protect, and some seem quite willing to sacrifice the odd swimmer (or the whole swim team) if they think there is a good profit to be made in the process. That's what this book is about: lousy lifeguards—people whose lack of training, conflicts of interest, or general disregard have put us all at risk of storming off the cliff like so many apocryphal lemmings.

I'm not saying that all of the lousy lifeguards are evil or ill-intentioned, although some may shake your faith in humanity. Rather, the whole lifeguarding institution seems to be failing, and not necessarily by accident. In the past two decades, and particularly on the issue of climate change, there has been an attack on public trust and a corresponding collapse in the integrity of the public conversation. The great institutions of science and government seem to have lost their credibility, and the watchdogs in media have lost their focus. Here we are, standing on the most dangerous environmental precipice that the human race has ever encountered, and we suddenly have to take a fresh and frightening look at the lifeguards in our midst.

The view is not reassuring. Take, for example, the case of Freeman Dyson. Dyson is an incredibly impressive character, a physicist who many people believe should have been given a Nobel Prize for his early work in quantum field theory. Later in his career he also distinguished himself as a good writer with a talent for simplifying and popularizing science. His 1984 antinuclear analysis, *Weapons and Hope,* won a National Book Critics Circle Award. Dyson was always a contrarian, but at age eighty-five (he was born on December 15, 1923), he has become fully argumentative. He is, for example, an outspoken skeptic of many aspects of modern climate science, and he has become a popular expert among those who would like to ignore or deny the risks of global warming.

That's all well and good. It makes sense that skeptics would seek out other skeptics to try to bolster their—perhaps delusional but perhaps sincere—opinions about climate change. It's also entirely reasonable that Dyson should want to keep up his profile and keep commenting on issues of scientific interest. But it doesn't explain why, on March 25, 2009, the *New York Times Magazine* would have presented an eight-thousand-word cover story on Dyson, lauding him as "the Civil Heretic." Neither does it explain why the *Times*, certainly one of the most respected sources of journalistic information on the continent, sent a sportswriter (Nicholas Dawidoff) to write the story. No criticism of Dawidoff: he's a wonderful writer, the author of some particularly excellent baseball books. But it's reasonable to ask why the *Times* would choose someone with no expertise, no education, and no background in climate science to interview a man apparently dedicated to undermining public confidence in the majority view about the risks of global warming.

As a lifeguard, the last time Freeman Dyson went down to the bottom of the cliff to check on the rock pile was, well, never. He too has no background in climate science, having done no research whatever—ever—on atmospheric physics or on climate modelling. Even in theoretical physics, his area of expertise, his greatest contributions date to the late 1940s and early 1950s. So again, in a free society Dyson has every right to stand at the top of the cliff and shout, "Jump!" But it's reasonable to wonder why the *New York Times Magazine* would give him the soapbox, especially when most of the time the magazine pays relatively little attention to this, the most urgent environmental issue humankind has ever faced.

Here's another fairly current example: the *Globe and Mail*, Canada's answer to the *New York Times* and arguably the most influential newspaper north of the 49th parallel, carried an opinion piece on April 16, 2009, by Bjørn Lomborg, the famously self-described *Skeptical Environmentalist* (per the title of his best-selling 2001 book). Under the headline "Forget the Scary Eco-Crunch: This Earth is Enough," the article sets out to dismiss the concern that humans are currently consuming global resources at a pace that cannot be sustained.

Lomborg begins by criticizing the concept of an ecological footprint, in which scientists try to estimate actual human impact on the environment rather than counting only the land we cover with roads and houses. As Lomborg says, scientists working on behalf of the World Wildlife Foundation have calculated that when you add up all the land affected by human consumption habits—the land where we live, the land used to grow our food, the land that is destroyed by mining or polluted by industries that produce our consumables—"each American uses 9.4 hectares of the globe, each European 4.7 hectares, and those in low-income countries one hectare. Adding it all up, we collectively use 17.5 billion hectares. Unfortunately, there are only 13.4 billion hectares available. So, according to the WWF, we're already living beyond Earth's means, using around 30 percent too much."

Complaining that these calculations oversimplify the situation and don't factor in potential future changes, Lomborg goes on to say, "... it is clear that areas we use for roads cannot be used for growing food, and that using areas to build our houses takes away from forests. This part of the ecological footprint is a convenient measure of our literal footprint on Earth. Here, we live far inside the available area, using some 60 percent of the world's available space, and this proportion is likely to drop because the rate at which the Earth's population is increasing is now slowing, while technological progress continues. So no ecological collapse."

This logic is impenetrable. Lomborg implies, first of all, that we can disregard the ecological aspect of our footprint because it's tricky to tally with absolute certainty. Then he says our literal footprint is actually going to get smaller because the population is rising, but at a slightly reduced rate. (Lomborg alone understands how more humans will take up less space.) Then, the skeptical environmentalist reassures us with this: "Due to technology, the individual demand on the planet has already dropped 35 percent over the past half-decade, and the collective requirement will reach its upper limit before 2020 without any overdraft."

That's wonderful, or it would be if it could be proven. But if Lomborg has some secret source of information for this contention, he is not sharing it with readers. Instead, he throws these assertions out without attribution or substantiation. He runs to the cliff, grabs the *Globe and Mail* megaphone, and shouts, "Jump!"

Again, that is his right. But why is Canada's leading newspaper promoting this as a reliable viewpoint? Lomborg is not a scientist (his Ph.D., in *political* science, concentrated on game theory), and his previous work has been widely and publicly criticized for its inaccuracy. (See Chapter 10 for more on Lomborg's checkered track record.) Why, even under the guise of "opinion," would a serious newspaper present this unsourced and inexpert argument as worthy of public attention?

It's not as though the true state of the world's environment is a mystery—or that it is left unstudied by leading and highly qualified scientists. For example, a collection of 1,360 such experts completed the Millennium Ecosystem Assessment in 2005. Those scientists, all leaders in their fields, concluded that, "over the past 50 years, humans have changed ecosystems more rapidly and extensively than in any comparable period of time in human history, largely to meet rapidly growing demands for food, fresh water, timber, fiber and fuel. This has resulted in a substantial and largely irreversible loss in the diversity of life on Earth."

"Substantial and largely irreversible." That sounds more dramatic than Lomborg's reassuring promise of "no ecological collapse." The whole Millennium Ecosystem Assessment report suggests very specifically that humankind is destroying the environment at a frightening pace. We are burning down forests, trashing the ocean, and changing global climate in a way that is making it extremely difficult for other species to survive—substantial and irreversible. In a way, we have to hope that Lomborg is right: we have to hope that this Earth *is* enough—and it may be, especially if humans pay attention to the warning signs and start behaving differently. But Lomborg is mounting a transparently fatuous argument to convince us that we don't have to pay attention to our ecological footprint. While more than thirteen hundred of the world's leading scientists try in good faith to back us away from the cliff, Lomborg grabs a soiled lifeguard T-shirt from a bin at the nearest thrift shop and tells us to keep jumping, ignore the risks. And the *Globe and Mail* cheers him on.

A third story broke in the early spring of 2009 that cast light on the weakness of modern lifeguard recruitment. On April 23, 2009, the *New York Times*'s excellent science writer Andrew Revkin reported on a now-defunct organization called the Global Climate Coalition, primarily a group of companies whose operations or products are heavy producers of greenhouse gases. For more than a decade, ending in 2002, the coalition spent millions of dollars on advertising and lobbying campaigns aimed at convincing public officials specifically and the public generally that climate change was not proven and that mitigating action was unnecessary. Yet, as Revkin reported, recently released court documents show that the Global Climate Coalition's own scientists had said in their 1995 report *Predicting Future Climate Change,* "The scientific basis for the Greenhouse Effect and the potential impact of human emissions of greenhouse gases such as CO_2 on climate is well established and cannot be denied."

It seems clear from the record that the Global Climate Coalition wasn't really interested in the science of climate change. Revkin reports that someone within the organization deleted the above reference and, even then, never distributed the report. And the group didn't actually invest in any climate change research. Instead it spent a fortune (the 1997 budget alone amounted to US$1.68 million) sowing confusion and lobbying against climate change policies, a gesture that, coincidentally or not, would serve the financial interests of the coalition's major funders: ExxonMobil, Royal Dutch Shell, British Petroleum (now BP), Texaco, General Motors, Ford, DaimlerChrysler, the Aluminum Association, the National Association of Manufacturers, the American Petroleum Institute, and others.

To take the crowded-cliff analogy one step further, it was as if some of the lifeguards had been charging thrill-seekers money to jump into the water, and they didn't want to give up the income. Not only did they pass up the opportunity to check the rocky bottom themselves, but when they hired someone to check, and that someone (in this case, a Mobil Corporation chemical engineer and climate expert named Leonard S. Bernstein) came back and said there was trouble below, they buried the report—and kept selling tickets.

You will in the coming pages meet a cast of lifeguards that in some instances may shake your faith in humanity. You will read about industry associations (the Western Fuels Association, the American Petroleum Institute) that commissioned strategy documents aimed at confusing people about climate science. You will see specific efforts to deny the gathering consensus that humans are endangering the planet—and you'll see how a group of think tanks and political operatives helped to implement the strategy, polluting the public conversation in North America and, increasingly, in Europe as well. You will read about "scientists" who strayed casually outside their field of expertise and then collected guest-speaker fees for also denying the advanced state of climate science understanding. You'll see a matter of well-established science skillfully recast as a subject for debate, as something that was primarily and hotly political and—until the intervention of admirable Republican leaders like John McCain and Arnold Schwarzenegger—destructively partisan. You will read about lobbyists like Steven "The

Junkman" Milloy, who took money from companies like Philip Morris, Monsanto, and ExxonMobil and then promoted himself as an expert commentator. Perhaps worst of all, you will see the great (and sometimes not-so-great) journalistic bastions of free speech employ or feature Milloy and others like him without ever telling the audience about the strained credentials or the conflicts of interest that might have affected the credibility of these wannabe lifeguards.

You may conclude from all this that reputable newspapers and magazines are today acting in a confused and confusing manner because a great number of people have worked very hard and spent a great deal of money in an effort to establish and spread that confusion. You will also see that their efforts have been disastrously successful. We have lost two decades—two critical decades—during which we could have taken action on climate change but didn't, because we were relying on bad advice. We were listening to lifeguards whose primary agenda had nothing to do with protecting our safety.

It's possible that when you see the full extent of the sometimes strategic, sometimes accidental campaign of confusion, you will drift into irritation, even into anger. You may want to blame the bad advisors—the freelance lifeguards whose real goal was often something other than swimmer safety. You may, especially, lose faith in mainstream media as a reliable source of credible information. After all, we rely on them for their judgment as well as for the accuracy of what they present in their newspapers and broadcasts, and on so many occasions they have let us down.

Finally, you might begin to lose hope. You might come to question our ability to have a credible public conversation about science and to arrive at a reasonable set of policies to address climate change. You might be tempted to throw up your hands in despair.

That would be the worst possible result. Just by picking up this book, you have made the first, critical step toward being part of the solution. The information that follows will at least help to inoculate you against the public relations spin, the confusion and misinformation that has led us through two decades of inaction. At best, it will inspire you to learn more about climate change and more about the practical, affordable, and essential things that we all need to do to conquer the problem.

Our species has proved itself capable of great stupidity and palpable evil. Human history is too full of pogroms and holocausts, of wars, genocides, and societal collapses. Equally, however, we have proved ourselves intelligent and adaptable. When we stepped back from the brink of global nuclear annihilation, we showed that when the conversation is open and accurate, we can make good, even altruistic decisions. It's time for such a decision now. It's time for good people to inform themselves, to help lead and guide their families, their friends, and their neighbors back from a path that threatens the habitability of planet Earth to one that will be sensible and sustainable. We don't have to jump off the cliff, and if someone tells you that we do, the message of this book is this: check his credentials. You may be surprised (and disappointed) by what you find.

[*two*]

THE INCONVENIENT TRUTH

Who says climate change is a scientific certainty?

No one, really. Certainties are rare in science. Even the reappearance of the sun over the horizon tomorrow morning can be reduced to a question of probability. On the question of climate change, scientists say they are more than 90 percent sure that it's happening and that humans are responsible, but you just never know.

Scientists embrace that kind of skepticism. It is through doubting the certainties of the world (the flatness of the Earth, the usefulness of bloodletting) that scientists advance human knowledge. But no serious scientist will stand up and denounce a widely accepted scientific theory without making a verifiable argument to the contrary. Scientists—real scientists—bind themselves to a strict discipline, setting out their theories and experiments carefully, subjecting them to review by other credible scholars who are knowledgeable in their field, and publishing them in reputable journals, such as *Science* and *Nature*.

The people who approach the science of climate change with that kind of integrity have agreed on its underlying components for years. The greenhouse effect, by which gases such as carbon dioxide absorb heat, setting up a warming blanket around the world, was first postulated by the French mathematician and physicist Joseph Fourier in 1824. Fourier understood that solar energy heated the Earth, which then reflected that heat back into space in the form of infrared radiation. In effect, the sun's heat bounced off the Earth's surface. But Earth's atmosphere seemed to be blocking or slowing the release of that infrared energy, warming the planet. In the 1850s the Irish physicist John Tyndall figured out a way to actually test and measure the capacity of various gases, including nitrogen, oxygen, water vapour, carbon dioxide, and ozone, to absorb and transmit radiant energy. By 1858 he had effectively proved Fourier's theory.

At the end of the century—the 19th century—the Swedish scientist Svante Arrhenius advanced the theory even further. Arrhenius, who is considered the founder of physical chemistry, was the first person to predict that humans might actually increase the temperature of the Earth by burning fossil fuels and, in the process, increasing the amount of carbon dioxide in the atmosphere. Fossil fuels themselves represent millions of years of stored carbon. Every living thing on Earth is composed of carbon in one form or another. Plants inhale carbon dioxide, which comprises one molecule of carbon and two of oxygen, then convert the carbon to carbohydrates and release the oxygen back into the atmosphere. Animals eat the plants. Over hundreds of millions of years these plants and animals have fallen dead into swamps or drifted lifeless to the bottom of the ocean, there to be covered up by layers and layers of other carboniferous matter. Under the right conditions—heat and pressure—those massive carbon piles have been converted to coal, oil, or natural gas. And over the last two centuries humans have been digging up those fossil fuels and setting fire to them, reintroducing the carbon to oxygen and releasing the resulting carbon dioxide back into the atmosphere. When Arrhenius considered the effect of this trend, he tried to calculate the effect of that increased carbon dioxide. He estimated that a doubling of atmospheric carbon dioxide would increase Earth's temperature by 3.8 degrees Fahrenheit. This was a stunning bit of science for the time, given that the most recent report of the Intergovernmental Panel on Climate Change estimates that a doubling of carbon dioxide will increase the global average temperature by between 3.6 and 8.1 degrees Fahrenheit. It's also unnerving, in that the concentration of carbon dioxide in the atmosphere has risen since 1850 by more than one-third, from 280 parts per million (ppm) to 385 ppm, and we are on track to hit Arrhenius's feared doubling by sometime near the middle of this century.

The next scientist to ring the climate change alarm was the American oceanographer Roger Revelle, the man who explained the greenhouse effect to former U.S. vice president and Nobel laureate Al Gore when both men were at Harvard in the late 1960s. In 1957 Revelle published a paper with the chemist Hans Suess in which they predicted a global warming. At the time Revelle suggested that such warming might even be a good thing, but he and Suess prescribed caution in their paper, saying that humans were conducting "a great geophysical experiment" with almost no conception of the consequences.

Befitting a society in which scientific understanding guides important social decisions, concern about this issue began to crop up in the political sphere as early as the 1960s. Then-president Lyndon Johnson said in a special message to Congress in February 1965 that "this generation has altered the composition of the atmosphere on a global scale through . . . a steady increase in carbon dioxide from the burning of fossil fuels."

By the late 1970s scientists were beginning to get twitchy, starting to speak with one increasingly concerned voice. A National Academy of Sciences report authored in 1979 by the scientist Jule Charney said, "A plethora of studies from diverse sources indicates a consensus that climate changes will result from man's combustion of fossil fuels and changes in land use." It also was becoming apparent that global warming was not as benign as it sounded. Scientists were beginning to understand that even a small increase in global average temperature could throw off a balance that had existed in Earth's climate since long before the time of humans. They began warning of melting glaciers and collapsing ice caps, of floods and droughts and rising tides. They began to contemplate a change in world living conditions that was more dramatic than anything in human history and more sudden than anything that had happened in hundreds of thousands of years.

The American political establishment joined the discussion in 1988, led by presidential candidate George H.W. Bush. Running against Democratic contender Michael Dukakis, then–vice president Bush said, "Those who think we are powerless to do anything about the greenhouse effect forget about the 'White House effect'; as president, I intend to do something about it." Bush promised, if elected, to convene an international conference on the environment: "We will talk about global warming and we will act."[1]

The newly elected president was, at first, as good as his word. Later the same year, after the world community gathered to create the Intergovernmental Panel on Climate Change (IPCC), Bush signed into law the National Energy Policy Act "to establish a national energy policy that will quickly reduce the generation of carbon dioxide and trace gases as quickly as is feasible in order to slow the pace and degree of atmospheric warming . . . to protect the global environment."

I offer all of the foregoing for context. I am neither a scientist nor a historian, and I have no intention in this book of jumping into the actual science "debate." For an in-depth overview, you can go online and read the Fourth Assessment Report of the IPCC, a scientific collaboration of unprecedented breadth, depth, and reputation. You can google Elizabeth Kolbert's brilliant *New Yorker* series, *The Climate of Man*. Or you can pick up one of the great populist science books on the subject: Canadian scientist Andrew Weaver's *Keeping Our Cool;* Australian scientist Tim Flannery's *The Weathermakers;* Kolbert's later book *Field Notes from a Catastrophe;* or Al Gore's book version of *An Inconvenient Truth.* Any one of these will give you a solid enough grasp of the science to leave you nervous about the state of our world.

My point, however, is that no one seemed to be confused about climate change in 1988. The great scientific bodies of the world were concerned, and the foremost political leaders were engaged. So what happened between then and now?

Well, here's what happened in science: with each new experiment, with each new report of the IPCC, with each new article published in legitimate peer-reviewed scientific journals, the science community became more certain that they were on the right track. Naomi Oreskes, a professor of history and science studies at the University of California, San Diego, tested that question in a paper she published in the journal *Science* in 2005. Oreskes searched the exhaustive ISI Web of Knowledge for refereed scientific journal articles on global climate change that were published between 1993 and 2003, and she analyzed them on the basis of whether they supported, contradicted, or took no position on the consensus that the human release of greenhouse gases was causing climate change. She found 928 articles—and not a single one took exception with the consensus position.

Clear enough. But what was happening in the mainstream media during the same period? The best answer to that question comes from the brothers Jules and Max Boykoff, who published an article in the peer-reviewed *Journal of Environmental Change* in 2003 titled "Balance as Bias: Global Warming and the U.S. Prestige Press." The brothers had searched the libraries of four "prestige" dailies in the United States—the *New York Times*, the *Wall Street Journal*, the *Washington Post*, and the *Los Angeles Times*—and had analyzed their coverage of climate change between 1998 and 2002. They found that while the scientific press was coming down 928 to zero in accepting or, at the very least, not denying climate change, in 53 percent of their stories these four newspapers quoted a scientist on "one side" of the issue and a spokesperson on the other. I say spokesperson rather than scientist for two reasons. First, the deniers were very often not scientists, but rather political ideologues or self-appointed "experts" from think tanks. Second, even when the experts had scientific credentials, in most cases those credentials were not relevant to the topic at hand. The experts were geologists or economists commenting outside their field of expertise, not climate scientists reporting on up-to-date peer-reviewed science.

Boykoff and Boykoff telegraphed their point about the mainstream media in the title of their paper "Balance as Bias." Journalists in the modern age find it all but impossible to stay up to speed on every issue, especially every issue of science. To protect themselves, they very frequently fall back on the notion of balance: they interview one person on one side of an issue and one person on the other. There is even a fairly common conceit in North American newsrooms that if both sides wind up angry about the coverage, the reporter in question probably got the story about right.

This has a degree of legitimacy when the subject matter is political, economic, or even moral. There are legitimate differences of opinion on the correct way to handle many political issues, and few economists agree on the right response to a specific economic event. And on a highly emotional issue such as abortion—one in which people are just as likely to be bringing forth points that are based in religion as they are to be talking about science—it is completely appropriate to canvas a range of opinions.

But science is a discipline in which there are legitimate subject experts, people whose knowledge is weighed and measured by their scientific peers. This is the process people use to decide, for example, on a new surgical method or on the structural strength of a new metal alloy. If a doctor recommended that you undergo an innovative new surgical procedure, you might seek a second opinion, but you'd probably ask another surgeon. You wouldn't check with your local carpenter, and you certainly wouldn't ask a representative of the drug company whose product would be rendered irrelevant if you had the operation. If you were building an apartment block or a bridge and someone offered a "state-of-the-art" new girder that was lighter and cheaper than the conventional alternative, you wouldn't accept the recommendation on the basis of the salesman's promises or even on the latest feature in *Reader's Digest*. You would insist on a testimonial from scientific sources.

That's not what's been happening in the public conversation about global warming. For most of the last two decades, while scientists were growing more convinced about the proof and more concerned about the risks of climate change, members of the general public were drifting into confusion, led there by conflicting stories that minimized the state of the problem and exaggerated the cost of solutions. Somehow, we have been spun.

[three]

FROM BERNAYS TO TODAY

A brief history of private prophets and public lies

SPIN DOCTOR: *(noun) A person employed to gloss over a poor public image or present it in a better light in business and politics, especially after unfavorable results have been achieved. A lobbyist; a* PR *person.*
WIKTIONARY, FEBRUARY 16, 2009

I have never liked the term "spin doctor," and I hate this definition—at least I hate that someone would propose "PR person" as a reasonable synonym. Public relations is not by definition "spin." Public relations is the art of building good relationships. You do that most effectively by earning trust and goodwill among those who are important to you and your business. And in more than thirty years of public relations practice, I have learned that the best way to achieve those goals is to act with integrity and honesty and to make sure everybody knows you are doing so.

Spin is to public relations what manipulation is to interpersonal communications. It's a diversion whose primary effect is ultimately to undermine the central goal of building trust and nurturing a good relationship.

Of course, lies are darned handy when the truth is something you dare not admit. Earning trust and goodwill is a nonstarter if you're a cigarette company peddling a product (often to children) that everyone knows is offensive, addictive, and potentially deadly. An impartial observer might come to the same conclusion about the fossil fuel industry. ExxonMobil doesn't really have to worry about its public image: because it has a stranglehold on a commodity that is also addictive (we need that energy to make our current economy function) and, in the current circumstances, ultimately life-threatening—especially for all those people who will not be able to adapt to dramatic changes in world climate. So when Exxon gives money to think tanks in support of programs that sow confusion about global warming, that isn't public relations. It's not an effort to build or maintain the quality of Exxon's reputation. It is, rather, a direct interference in the public conversation in a way that serves Exxon's interest at the expense of the public interest.

But here's the part that bugs me the most: the people who are taking Exxon's money are often *in* public relations. Or they are taking advantage of skills, tactics, and techniques that have been developed and refined in the shadier parts of the public relations industry. Just as there are unscrupulous lawyers who use their expertise to help break the law, or unprincipled accountants who help their clients evade taxes, it seems there have always been public relations people willing to meddle with the public discourse to promote the private interests of the people who are paying their bills.

The Public Relations Society of America has a professional code of ethics, which begins: "I pledge to conduct myself professionally, with truth, accuracy, fairness, and responsibility to the public." I would urge you to keep that pledge in mind and to measure the stories that unfold in the following pages against that standard. I'm pretty sure you'll be disappointed.

But I am equally convinced that it is important for you to hear these stories, to learn about how sometimes-questionable public relations tactics have evolved, and to arm yourself against the effect of those tactics in the future. It's just as Aristotle said more than two thousand years ago: someone who is highly trained in rhetoric can argue any question from every angle—a skill that can be used for good or ill. But Aristotle didn't teach rhetoric so shysters could play the public for fools. Rather, he was trying to make sure that people would recognize when someone was playing with the language rather than promoting the truth. He taught rhetoric to inoculate the public against that kind of abuse.

Looking back into the history of public relations can be inspiring, but, I have to admit, it can also be disillusioning. Consider the examples of Ivy Lee (1877–1934) and Edward Bernays (1891–1995), two men who, perhaps unfortunately, will forever compete for the title of "the father of public relations."

If I could look only at what Ivy Lee said and disregard a lot of what he did, he'd be an appropriate hero. I say that because I see in Lee's own writings some of the same advice that I am in the habit of giving myself. For example, when a client asks me the key to establishing, maintaining, or recovering a good reputation, I say three things:

1. Do the right thing;
2. Be seen to be doing the right thing; and
3. Don't get #1 and #2 mixed up . . .
   . . . by which I mean, always make sure that you're doing the right thing for its own sake and not for the reputational advantage you might gain.

Ivy Lee's prescription sounds pretty similar. He said, "Set your house in order; then tell the public you have done so." Do the right thing; be seen to be doing the right thing. So far, so good. Lee is also famously said to have told John D. Rockefeller Jr., "Tell the truth, because sooner or later the public will find out anyway. And if the public doesn't like what you are doing, change your policies and bring them into line with what people want."

This is all excellent advice, especially appropriate if you are trying to recover your reputation after an unfortunate accident. In fact when I first started thinking this way, it wasn't because I was trying to force an ethical framework on the public relations business. It was because I had learned that this is what works. I had noticed that when my clients tried to cover up bad news or gloss over problems, those problems got worse. But when people stood up, told the truth, and did the right thing, they won public trust and earned higher regard.

It was obvious from early in his career that Lee also understood the importance of openness and integrity in building a good reputation. In 1906, for example, after a train crash on the Pennsylvania Railroad, he convinced management to forego the usual approach of bribing reporters to ignore the story in favor of throwing open the doors—actually bringing reporters to the scene at the railroad's expense and offering all the assistance they might need once they got there. It worked like a charm. People understand that accidents happen, and are remarkably forgiving, especially if they see you making an effort to "set your house in order" and prevent an unnecessary recurrence.

Lee also distinguished himself during the First World War by organizing publicity for the fledgling Red Cross. He is credited with helping that organization grow in the United States from 486,000 to 20 million members by the time the war was over. And he helped to raise US$400 million—an unbelievable fortune given the strength of the dollar of the day.

Like so many who came after him, however, Lee found that the high road is a less attractive option when you're managing a crisis that is not so easily forgiven. Working for the Rockefeller family in 1914, Lee was called in to manage media after the Colorado state militia and company guards had sprayed machine-gun fire into a colony of striking mine workers. The guards also set fire to the colony's tents, resulting in the deaths of twenty-two people, including eleven children. Three guards were also killed. Lee responded by launching a series of bulletins titled "The Struggle in Colorado for Industrial Freedom," demonizing the strikers and lionizing the supposed heroism of the Rockefeller guards. He did something similar in West Virginia, where seventy coal miners lost their lives in labor disputes. This time the bulletin was called "Coal Facts," but the "facts" that Lee selected were particularly favorable to the Rockefellers.

Far from causing him a professional problem, these adventures helped seal Lee's reputation as one of the most prominent and successful early public relations practitioners—a reputation that he lost rather abruptly in 1933 when he went to work for the German Dye Trust, trying to promote relations between the United States and Nazi Germany and leaving a stain that still blots his name.

Lee's big competitor for the title "father of public relations" is Edward Bernays, an Austrian immigrant to the United States and Sigmund Freud's nephew. Bernays is the person who coined the term "engineering consent." In his 1928 book *The Business of Propaganda,* Bernays put into words something that every demagogue in history probably knew instinctively. He wrote, "If we understand the mechanism and motives of the group mind, is it not possible to control and regiment the masses according to our will without their knowing about it? The recent practice of propaganda has proved that it is possible, at least up to a certain point and within certain limits."

It would be foolish to take issue with this or even to insist that it is necessarily evil in its conception or intent. As anyone who has ever been involved in a public relations or political campaign knows, humans are not coldly rational in their decision making or uniformly predictable in their responses. They are generally busy, a little harried, and appropriately obsessed with their own affairs. In most cases it's hard to get their attention at all, and when you do, people often respond emotionally or on the strength of biases that may be a complete mystery to an advertiser, an advocate, or a politician hoping to engage in a straight conversation. Effective public relations requires a degree of subtlety. You have to make the effort to understand why people think the way they do, and you have to find a way to communicate in a fashion that will enable them to understand you.

Bernays also recognized both the benefits and dangers of developing and using such skills. Writing in his other 1928 book, *Propaganda*, he said that a public relations counsel (another term he appears to have coined) "must never accept a retainer or assume a position which puts his duty to the groups he represents above his duty to society."

Bernays then built a great career faithfully exercising his duty to his clients in a way that often seemed to disadvantage society. Some of this was relatively harmless; for example, he organized the first known political pancake breakfast (for Calvin Coolidge). He also organized the Torches of Liberty Brigade in Manhattan in 1929. In what was presented as a demonstration for women's equality, Bernays assembled a crowd of young women who marched in that year's Easter parade smoking Lucky Strikes, asserting their right to smoke in public. This stirring performance was paid for by a relatively small investment from the American Tobacco Company, which got the benefit when women felt "liberated" enough to start smoking in public.

If you consider how little was known at the time about the dangers of smoking, you might be able to pass this off as a cute and clever campaign. (Bernays himself said before his death in 1995 that he would never have organized the event if he had known that smoking was to become one of the principal health threats of the century.) Yet even today, the "torches" parade is used in public relations courses across the country as an example of how you can earn free media attention and shift the public view of an issue in an indirect way. In the way people enjoy being fooled by a good magician, they seem willing to forgive Bernays for having tricked them with a public relations event that at the time he would have argued was harmless.

Less forgivable was Bernays's participation in the campaign (and ultimate CIA coup) to oust the democratically elected government of Guatemala in 1954, an incident that put the interests of the United Fruit Company (now Chiquita Brands International) ahead of all others. Bernays also noted in his own autobiography that Nazi propagandist Joseph Goebbels praised another of Bernays's books, *Crystalizing Public Opinion,* as having been helpful in crafting the campaign against German Jews.

It would not be fair or accurate to draw some kind of Nazi propagandist thread from Ivy Lee and Edward Bernays through a whole century of public relations abuses and tie all of that to the campaign to confuse people about climate change. But it might be worth contemplating the slippery slope that faces people in public relations who forget their duty to society—the Public Relations Society of America's caution to practice "professionally, with truth, accuracy, fairness, and responsibility to the public."

In an adversarial world full of lawyers, where you get used to hearing one person on one side of an issue and one on the other, a danger exists that public relations people will begin to think of themselves not as communicators with a responsibility to their audience but only as advocates. In court (and before you conclude that I am lawyer-bashing, I learned all this in law school myself), there is a convention that every accused person deserves the best possible defense, and it is the lawyer's duty to mount that defense to the best of his or her ability. We have even grown to accept the idea that it's acceptable to construct a case that is entirely—almost deceptively—one-sided, knowing that the lawyer on the other side will bring equal vigor to the case. This approach appears to have carried over to public relations and to the court of public opinion. Some public relations people act as though it is their duty to mount the most compelling—or most devastating—case possible on behalf of their client, leaving it to the opposition to mount a counter argument, and allowing the public to sort it out.

The problem is this: in court, there are rules of evidence (you have to tell the truth) and a judge who has the expertise and is given the resources to make an intelligent decision about what is being presented. But there are no such rules in the public conversation. There are only tactics, strategies, and spin.

Although it is not always successful in doing so, the court also endeavors to level the playing field when one party is rich and powerful and another is pressed for resources. In the court of public opinion, however, there is no such corrective. Rich individuals, large corporations, and industry associations can afford to muster a devastating campaign, against which environmentalists or conscientious scientists must always strain to respond.

At the end of the day, it comes back to the rules of ethical practice. As Edward Bernays might have said, it's okay to put lipstick on a vice president (or a vice-presidential candidate), but you should always call a pit bull a pit bull. That's not what's been happening in the climate change conversation. A public policy dialogue that should have been driven by science has instead been disrupted by public relations—and if you look closely, it seems to be the kind of public relations that Ivy Lee and Edward Bernays practiced on their worst days, not the kind they recommended on their best.

[*four*]

THE AGE OF ASTROTURFING

In which industry steals credibility from the people

It was a conspiracy!

There's something histrionic about that charge. The very idea of a cabal of rich and powerful people conspiring to fool the public about a fundamental point of science strains credulity and is offensive in its own right. Yet if you read on, you will see that there are conspiracies aplenty, documented and undeniable.

The first was organized by the Western Fuels Association, which as of April 2009 defined itself on www.westernfuels.org as "a not-for-profit cooperative that supplies coal and transportation services to consumer-owned electric utilities throughout the Great Plains, Rocky Mountain and Southwest regions." The magic word in that description is "coal," the most plentiful conventional energy source in the world and the number-one fuel for electric utilities in the United States, which has the second-largest known deposit of coal in the world, only slightly behind Australia. The problem is that coal is also the worst fossil fuel when it comes to generating carbon dioxide, and those coal-fired electrical generators are already the largest carbon dioxide point source in the country.

In 1991 Western Fuels joined with the National Coal Association and the Edison Electric Institute to create the Information Council on the Environment (ICE). This was a not-very-arm's-length organization that would use its original US$500,000 budget "to reposition global warming as a theory (not fact)" and "supply alternative facts to support the suggestion that global warming will be good."[1]

ICE went into small U.S. markets that were heavily dependent on coal-fired electricity and, with advance planning from the D.C. public relations firm Bracy Williams and Company, tested a series of messages, including:

- "Some say the Earth is warming. Some also said the Earth was flat."
- "Who told you the Earth was warming . . . Chicken Little?"
- "How much are you willing to pay to solve a problem that may not exist?"

It actually wasn't getting warmer in Minneapolis, and presumably the messaging went down well, especially on cold winter days, because ICE rolled out a campaign that included newspaper and radio advertising. ICE also learned that audiences didn't take coal or electrical company officials very seriously when it came to arguing environmental issues, but that they were inclined to listen to "technical experts." So ICE mobilized a group of scientists who in many instances were not climate change experts, but who would nevertheless make themselves available for newspaper and broadcast interviews and sign opinion page articles that could be distributed to local papers.

Parallel to the ICE operation, the Western Fuels Association also launched another "educational" entity called the Greening Earth Society, which produced a video called *The Greening of Planet Earth*, a thirty-minute love note to carbon dioxide that is still available for viewing on YouTube. This became the first public appearance of a group of scientific experts made up of people like Sherwood Idso—people who have since become famous for their willingness to argue climate science on behalf of the fossil fuel lobby. In the video they argue that Earth's plants are starving for carbon dioxide and that an increase in atmospheric carbon dioxide will result in a more fertile world. Ignoring the implications of climate change, especially the threat of lasting droughts that could turn much of the equatorial zone into a desert, *The Greening of Planet Earth* showed a time-lapse animation in which carbon dioxide–driven vegetation colonizes virtually every part of the Earth's surface—even closing in happily over the Sahara. The message was clear: climate change—if it's happening—is a good thing.

The Western Fuels Association offered the video online in return for a small, tax-deductible donation to the Greening Earth Society, but it delivered hundreds of copies for free to public and university libraries across the country. As Naomi Oreskes reports in her fabulous podcast at smartenergyshow.com, "You CAN Argue with the Facts," the overworked librarians at the University of Oregon took this gift at face value, filing it with the description that the Western Fuels Association had provided: "An enlightening documentary that examines one of the most misunderstood environmental phenomena of the 1980s." Imagine the potential confusion to be suffered by a first-year student who has been reading legitimate science about global warming and checks this video out of his university library, in all probability becoming the first person at the institution to actually watch it. On one hand, the student would have learned in class that climate change was a gathering threat. On the other, the university was inadvertently endorsing a contrary argument that global warming would be a boon to humanity.

The Western Fuels Association put ICE on ice after one of its strategy documents was leaked to the newspapers, sparking a raft of embarrassing stories in the *Energy Daily*, the *National Journal*, the *Arizona Republic*, and the *New York Times*. But a pattern was beginning to take hold. Corporations and industry associations were using their considerable financial resources to influence the public conversation. They were using advertising slogans and messages that they had tested for effectiveness but not for accuracy. They were hiring scientists who were prepared to say in public things that they could not get printed in the peer-reviewed scientific press. And they were taking advantage of mainstream journalists' willingness—even eagerness—to feature contrarian and controversial science stories, regardless of whether the controversy was actually occurring in reputable scientific publications.

The next example of a transparent effort to manipulate public opinion on a range of issues, including climate change, started out as a project of the tobacco giant Philip Morris. Big Tobacco had been playing this game since the days of Bernays, at first trying to surround cigarettes with a patina of glamor and then wrapping the death sticks in a cocoon of doubt. It began with the founding of the Tobacco Institute in the 1950s and specifically with the creation of the Tobacco Industry Research Committee, later the Council for Tobacco Research. The Tobacco Institute and the Council for Tobacco Research were both tireless in funding and promoting any research that would cast doubt on the health effects of smoking. There is a great scene in the 2005 movie *Thank You for Smoking* in which the main character, Nick Naylor (played by Aaron Eckhart), talks admiringly about a cigarette industry scientist who had done research on tobacco for thirty years without finding a link to cancer. About which Naylor says, sardonically, "The man's a genius."

From the emergence of the tobacco lobby in the 1950s until the tobacco companies started losing huge health-related lawsuits in the 1990s, the tobacco industry's message was admirable for its consistency: the link to cancer (and, later, the cancer link to secondhand smoke) was not "proven." Tobacco defenders said the alleged link was based on epidemiological studies that established a correlation but couldn't prove cause and effect beyond a reasonable doubt. They also made arguments that seemed calculated to distract people from the actual issue. They said that lots of things caused cancer, so it was unreasonable to try to pin all lung cancer deaths on tobacco or to pick on cigarettes and not deal with all the other causes at the same time. And they criticized as zealots anyone who tried to educate or legislate against tobacco use, saying that the health advocates, government bureaucrats, or responsible politicians were creating a nanny state that would interfere with people's rights.

This was a highly effective mixed-message strategy. The smoky executives knew they were never going to win the health argument, so they muddied the scientific waters and tried to reposition the debate to be about free choice. According to a document obtained by the organization TobaccoFreedom.org,[2] the executives even co-opted the American Civil Liberties Union, providing big donations to the ACLU in return for its support in recasting smoking as a matter of freedom and individual choice.

Still, by the late 1980s the public had grown tired of tobacco industry "geniuses" telling them smoking was harmless, and skeptical of tobacco company employees or institute "experts" fighting against increasingly popular smoking restrictions. So Philip Morris opened up two new fronts. First, working with the public relations giant Burson-Marsteller, Philip Morris financed the creation of the National Smokers Alliance, a purported grassroots association that mustered smokers together to fight for their "rights."

From a tactical standpoint this was a brilliant strategy. True grassroots organizations are one of the great expressions of democracy. In them, theoretically at least, you have a group of independent citizens bound together in common interest rising up and demanding to be heard. Reporters have grown appropriately cynical of corporate manipulation, and they are generally suspicious of established interest groups, from environmentalists to consumer advocates, who have made what appears to be common cause with government regulators. They find it refreshing to see an apparently spontaneous outpouring of support or opposition on any public issue.

The political theorist Jeffrey Berry documented this in his 2000 book, *The New Liberalism: The Rising Power of Citizen Groups.* Berry showed that in the realms of politics and media, grassroots organizations were outperforming industry-sponsored interest groups by a wide margin. For example, Berry found that a small number of citizens' groups making representations in Congress were overrepresented in media citations by a factor of 10 to 1 when compared to their industry counterparts.

The National Smokers Alliance, however, was not a spontaneous outpouring of public support. It was an Astroturf group, a fake grassroots organization animated by a clever public relations campaign and a huge budget. As John Stauber wrote in a 1994 edition of the online journal PR *Watch:*

> Burson-Marsteller's state-of-the-art campaign utilizes full-page newspaper ads, direct telemarketing, paid canvassers, free 800 numbers, newsletters and letters to send to federal agencies. B-M is targeting the fifty million Americans who smoke. Its goal is to rile-up and mobilize a committed cadre

> of hundreds of thousands, better yet millions, to be foot soldiers in a grassroots army directed by Philip Morris's political operatives at Burson-Marsteller.[3]

Such are the profits available to the tobacco industry that around the same time, Philip Morris had also engaged the public relations giant APCO Worldwide to craft a parallel attack on the scientific validity of links between cancer and second-hand smoke. In 1993 APCO proposed the foundation of The Advancement of Sound Science Coalition (TASSC). The original documentation of this proposal is available on the Internet at TobaccoDocuments.org, a Web site established after the tobacco industry lost a series of 1990s lawsuits over the falsification of evidence and the attempt to cover up the health effects of smoking.

The Advancement of Sound Science Coalition's stated objectives were these:

- Establish TASSC as a credible source for reporters when questioning the validity of scientific studies.
- Encourage the public to question—from the grassroots up—the validity of scientific studies.
- Mobilize support for TASSC through alliances with other organizations and third-party allies.
- Develop materials, including new article reprints, that can be "merchandized" to TASSC audiences.
- Increase membership in and funding of TASSC.
- Publicize and refine TASSC messages on an ongoing basis.[4]

The eagerness to increase TASSC's membership and funding is important in the climate change conversation because, well, guess who APCO contacted when it came time to increase membership in its new Astroturf group? Realizing how obvious it would look if Philip Morris was TASSC's only financial supporter, APCO sent out recruitment letters to twenty thousand businesses inviting them to join the fight for "sound science." Below is a list of the kinds of companies that APCO considered appropriate partners for such a venture. The list comes from another memo, written in 1994 when APCO was planning to expand TASSC operations into Europe.[5] One of the first goals, the memo said, was to try to "link the tobacco industry with other more 'politically correct' products."

"As a starting point," the memo begins, "we can identify key issues requiring sound scientific research and scientists that may have an interest in them." This seems to suggest that no one was currently conducting research in any of the areas about to be discussed, or that the research being done was somehow unsound. Certainly, that research in most cases seemed to be inspiring legitimate interest groups to demand more government intervention and regulation of the sort that was costing industry money.

The TASSC memo continued:

> Some issues our European colleagues suggest include:
>
> - Global warming
> - Nuclear waste disposal
> - Diseases and pests in agricultural products for transborder trade
> - Biotechnology
> - Eco-labeling for EC products
> - Food processing and packaging

It's worth looking at these tactics because they suggest a high degree of sophistication and a willingness to underwrite an ambitious and expensive "grassroots" campaign. First, APCO suggested that TASSC target secondary markets, because this would "avoid cynical reporters from major cities [and involved] less reviewing/challenging of TASSC messages." It's a judgment call whether big-city reporters are by definition cynical. But it's pretty clear that they are, as a group, better educated, better informed, and more likely to be briefed on specific areas of science. Small-town papers and broadcast outlets tend to have fewer journalists in total and fewer specialists. They also pay less, so there is an incentive for the best reporters to get their early experience in smaller markets and then move up to the high-paying big-city jobs. The public relations professionals at APCO know this, so you have to read more critically when the plan they wrote for TASSC recommends targeting small towns. The 1993 memo "The Revised Plan for the Public Launching of TASSC" suggests that a secondary-market focus "... increases [the] likelihood of pick-up by media" and "limit[s] potential for counterattack. The likely opponents of TASSC tend to concentrate their efforts in the top markets while skipping the secondary markets." It seems fair to conclude that this is APCO's way of saying that TASSC should try to stay under the radar.

APCO also recommended that TASSC establish a public information bureau that would brief "pertinent associations" in Washington, D.C., and coordinate with "organizations that have tangential goals to TASSC, such as... The Science and Environmental Policy Project." (The latter is another Astroturf group established by a tobacco-sponsored scientist named Dr. S. Fred Singer—more about him later.) The name "public information bureau" sounds benign, like an informational clearinghouse that perhaps would send out the odd news release or be at the ready to answer questions. Certainly APCO anticipated sending out information, but the specific list of proposed activities and tactics suggested something much more proactive—and much more political. The list included:

- Publishing and distributing a monthly update report for all TASSC members, which will quantify media impressions made the prior month and discuss new examples of unsound science.
- Monitoring the trade press (e.g., public interest group newsletters and activities) and informing TASSC members of any upcoming studies and relevant news.
- Arranging media tours.
- Issuing news releases on a regular basis to news wire services, members, allies and targeted reporters.
- Acting as a clearinghouse for speaking requests of TASSC scientists or other members and maintaining a Speakers Bureau to provide speakers for allies and interested groups.
- Drafting "boilerplate" speeches, press releases and op-eds [opinion page articles] to be used by TASSC field representatives.
- Placing articles/op-eds in trade publications to serve as a member recruitment tool in targeted industries, such as the agriculture, chemical, food additive and biotechnology fields.
- Monitoring the field and serving as a management central command for any crises that occur.
- Developing visual elements that help explain some of the issues behind unsound science.[6]

Here again, APCO was advising that TASSC look for every opportunity to attack inconvenient ("unsound") science. It wanted TASSC to identify "targeted reporters" who would be most likely to give the kind of coverage that served TASSC's purposes. It suggested the drafting of "boilerplate" speeches and press releases of the kind that could be used again and again to promote its messages. And the op-eds could come in especially handy, again, in the secondary markets. As any big public relations firm (or would-be thought leader) knows, it's hard as can be to get the *New York Times* to publish your opinion piece. But if you are satisfied to send lots of copies to lots of smaller papers, many of which find it much harder to source a steady supply of good material, you have a reasonable chance of reaching just as many readers.

TASSC's early membership list included "sound science" supporters like Amoco, Exxon, Occidental Petroleum, Santa Fe Pacific Gold Corporation, Procter & Gamble, the Louisiana Chemical Association, the National Pest Control Association, General Motors, 3M, Chevron, and Dow Chemical. For "science advisors," they had people such as Frederick Seitz.

In the 1960s and '70s, Seitz was a widely admired scientist, a former president of both the National Academy of Sciences and Rockefeller University. In 1978 he took that reputation to work for the R.J. Reynolds Tobacco Company. According to "While Washington Slept," a May 2006 *Vanity Fair* article by investigative journalist Mark Hertsgaard, over a ten-year period Seitz was responsible for handing out US$45 million in tobacco money to people who were pursuing research that overwhelmingly failed to link tobacco to anything the least bit negative. Seitz later admitted to accepting almost US$900,000 of that money himself.

But by the late 1980s he seemed to have lost a step. In a Philip Morris interoffice memo dated 1989, an executive named Alexander Holtzman reported that he was told that "Dr. Seitz is quite elderly and not sufficiently rational as to offer advice."[7] Yet Seitz continued to stand as a TASSC regular, in particular lending his name and leveraging his old National Academy of Sciences affiliation to the global warming denial movement for nineteen more years, dying in 2008 at the age of ninety-six.

It's probably time to introduce Steven Milloy, a.k.a. "The Junkman," to our cast of characters. While TASSC was originally run by executive director Garrey Carruthers, an economist and former New Mexico governor, Steven Milloy took over in 1997. Milloy's academic background is also considerable: he has an undergraduate degree in science from Johns Hopkins, a masters in health sciences and biostatistics (also from Johns Hopkins), and a masters in law from Georgetown. But there is no record of his directly pursuing science or law as a career.

Instead, Milloy emerged in the 1990s working for a series of public relations and lobby firms, including the EOP Group, which the Web site PR Watch.org describes as "a well-connected, Washington-based lobby firm whose clients have included the American Crop Protection Association (the chief trade association of the pesticide industry), the American Petroleum Institute, AT&T, the Business Roundtable, the Chlorine Chemistry Council, Dow Chemical Company, Edison Electric Institute (nuclear power), Fort Howard Corp. (a paper manufacturer), International Food Additives Council, Monsanto Co., National Mining Association, and the Nuclear Energy Institute."[8]

Milloy, who is currently an "adjunct scholar" at the Competitive Enterprise Institute and formerly held that position at the Cato Institute, is also the creator and proprietor of the Web site JunkScience.com, which works to "debunk" everything from the dangers of secondhand smoke to the risks of genetically modified foods. Milloy was a founding member of a team assembled by the American Petroleum Institute (API) to create a 1998 "Global Climate Science Communication Action Plan" (the precise contents of which Greenpeace later discovered and made available for public viewing).[9] The API made no bones about its intent in creating its plan for the public. The document plainly states that its purpose is to convince the public, through the media, that climate science is awash in uncertainty. Notwithstanding that the industry's own scientists were saying as early as 1995 that the science of climate change was undeniable (as in the *New York Times* report discussed in Chapter 1), the API set out an entire strategy bent on making doubt, in the words of the memo below, "conventional wisdom." The API document begins with a kind of mission statement (the parenthetical additions appear as in the original):

> Victory Will Be Achieved When
>
> - Average citizens "understand" (recognize) uncertainties in climate science; recognition of uncertainties becomes part of the "conventional wisdom"
> - Media "understands" (recognizes) uncertainties in climate science
> - Media coverage reflects balance on climate science and recognition of the validity of viewpoints that challenge the current "conventional wisdom"
> - Industry senior leadership understands uncertainties in climate science, making them stronger ambassadors to those who shape climate policy
> - Those promoting the Kyoto treaty on the basis of extant science appear to be out of touch with reality.

The statement seems to make clear that the goal was not to promote an understanding of science, but to spread uncertainty. The goal was not to put the best case before a deserving public, but to ensure at all times that the public was treated to "balance"—and in this case, the API strategists meant that every time a top scientist offered the public new insights into the risks of climate change, the institute would be there with a contradictory view. Victory would be achieved when the public accepted this balance—this confusion—as "conventional wisdom." It was also a priority that industry leaders learn not about science but about uncertainty, with a specific goal of attacking the Kyoto agreement, making its supporters appear "out of touch with reality." There is, however, no contention here that Kyoto supporters really *were* out of touch, only that the API would like to cast them as such.

The plan went on to describe how the API might achieve these goals, beginning with a campaign to search out and recruit "new (scientific) faces who will add their voices to those recognized scientists who already are vocal." The document goes on to expand on the list of specific tactics (with my emphasis added in italics):

- Develop a global climate science information kit for media including peer-reviewed papers that undercut the "conventional wisdom" on climate science. This kit also will include understandable communications, including simple fact sheets that present scientific uncertainties in language that the media and public can understand.*
- Conduct briefings by media-trained scientists for science writers in the top 20 media markets, using the information kits. Distribute the information kits to daily newspapers nationwide with offer of scientists to brief reporters at each paper. Develop, disseminate radio news releases featuring scientists nationwide, and offer scientists to appear on radio talk shows across the country.
- Produce, distribute *a steady stream of climate science information* via facsimile and e-mail to science writers around the country.
- Produce, distribute via syndicate and directly to newspapers nationwide *a steady stream of op-ed columns and letters to the editor authored by scientists.*

* For the record, and per the discussion in Chapter 2 about Naomi Oreskes's survey of the reputable scientific literature of the day, neither the API nor any other organization that wants to deny the science has yet come up with any "peer-reviewed papers that undercut the 'conventional wisdom' on climate science."

- Convince one of the major news national TV journalists (e.g., John Stossel) to produce a report examining the scientific underpinnings of the Kyoto treaty.
- Organize, promote and conduct through grassroots organizations a series of campus/community workshops/ debates on climate science in 10 most important states during the period mid-August through October, 1998.
- Consider advertising the scientific uncertainties in select markets to support national, regional and local (e.g., workshops / debates), as appropriate.

Like the Western Fuels Association campaign in the early 1990s and the TASSC campaign that followed, this document once again set out a major work plan that involved burying science writers in "a steady stream of climate science information" concentrating not on quality but on doubt. It can hardly be a coincidence that even as the science itself was becoming ever *more* certain—and ever more alarming—the "conventional wisdom" in the late 1990s and into the early part of this century turned more and more to confusion and doubt.

If you look at the bottom of the "Situation Analysis" within this plan, you get a list of the authors. The list includes but is not limited to Candace Crandall, Science and Environmental Policy Project; Jeffrey Salmon, George C. Marshall Institute and later the Bush Administration's associate under secretary for science at the U.S. Department of Energy; Myron Ebell, Frontiers of Freedom; Randy Randol, Exxon; Sharon Kneiss, Chevron; Steven Milloy, TASSC; and Joseph Walker, American Petroleum Institute. The more you read in this area, and the farther you read into this book, the more you begin to recognize the names of people and organizations. You also find that many of the most prominent "scientists" and spokespeople are not currently working in science, and often never were working in climate science. Many others, like Steven Milloy, enjoy generous financial connections to self-interested industries, connections that they generally fail to report when they are casting themselves as impartial experts.

Returning to the question of Astroturf groups, you also realize that the term "grassroots" as we might normally recognize it means something completely different to the people who are writing these reports. When we think of a grassroots group, we might think of something like the Montgomery Improvement Association, the citizens' group that emerged to support Rosa Parks when she stood up in 1955 for the rights of African Americans. But like any number of modern public relations firms that boast of having grassroots expertise, the API was talking about something much less spontaneous. The API's grassroots groups were not going to sprout up independently. They were to be planted, tended, nurtured, and financed by the fossil fuel companies that would benefit as the actual weight of science gave way to a manufactured "conventional wisdom."

There are four specific references to "grassroots" in the API document. First, the API proposed establishing a "global climate science data center," staffed not by scientists but by "professionals on loan from various companies and associations with a major interest in the climate issue." One of the important prerequisites for these "professionals" was that they have "expertise in grassroots organization." This "science data center" could then start a "national direct outreach and education" project, one element of which would be a plan to "distribute educational materials directly to schools and through grassroots organizations of climate science partners (companies, organizations that participate in this effort)." And again, from the earlier list of tactics, the API would "organize, promote and conduct through grassroots organizations a series of campus/community workshops/debates on climate science in 10 most important states during the period mid-August through October, 1998." In each of these strategy planks, the proposed "grassroots" groups do not currently exist but can be organized by people with the appropriate expertise. The result is not being pitched as a spontaneous expression of democratic choice, but as a fixed goal that can be achieved by patching together something that looks like a public organization built from the ground up, rather than an industry-driven lobby.

The other thing you'll notice if you sit down and read one of these documents is that the doubt about climate science begins to sound legitimate. You begin to forget that most of the "scientists" who act as spokespeople for the API or its partner organizations do *no* research in climatology or any other related field. You stop noticing that the goals of these "science" reports never include financing actual scientific research—or even an impartial review of the best of current science. The Global Climate Coalition asked its own in-house scientists for an impartial review in 1995, and then stuck the results in a drawer, far away from the curious eyes of the public.

No, promoting scientific research or advancing the public understanding of the true state of science appears not to be the priority. The API strategists, working on behalf of clients in the chemical and fossil fuel industries, are working instead to change the conventional wisdom, irrespective of the science. They are crafting a plan to create grassroots organizations that serve industry goals, regardless of whether the public might share those goals or might spontaneously have risen to fight for those priorities. These industry-funded planners set out to ridicule the Kyoto agreement and to frustrate government efforts to constrain greenhouse gas emissions. They made a plan to overwhelm the media with a steady stream of information that served industry's purposes and injected "balance" into coverage, whether or not that balance reflected the true state of science.

In March of 2009 Gallup updated its annual poll asking whether Americans thought the risks of climate change were being reported reasonably or whether they were being exaggerated. A total of 41 percent of respondents said they thought the seriousness of the global warming threat is, even at the beginning of 2009, still being exaggerated. That means that two years after the release of a report in which a Nobel Prize–winning group of scientists declared with a more than 90 percent certainty level that climate change is real, pressing, and apt to change forever the face of the Earth, four in ten Americans still don't believe it. For them, uncertainty is embedded as conventional wisdom. For the writers of the API strategy—though not for the defenders of accuracy in media—that must be considered a victory.

[*five*]

INTERNATIONALIZING UNCERTAINTY

Taking the doctrine of doubt on the road

There are few "skeptical scientists" with as little actual expertise and as much ambition as the Canadian geography professor Dr. Timothy Ball. Never a climate scientist per se, Dr. Ball quit his position as an associate professor at the University of Winnipeg in 1995, apparently ending an academic career that featured a lifetime output of just four peer-reviewed journal articles, none of which addressed atmospheric science. Yet ten years later, Ball-the-climate-expert seemed to be everywhere—on the radio, in the newspapers, on the lecture circuit, even testifying before a committee in the Canadian parliament. Online videos of his radio and lecture performances showed him to be an affable and entertaining speaker with a warm, funny, and folksy style. He would say things like this, quoted from an interview with Charles Montgomery for an August 12, 2006, story, "Meet Mr. Cool," in the *Globe and Mail:* "Environment Canada [the Canadian national weather service] can't even predict the weather! How can you tell me that they have any idea what it's going to be like 100 years from now if they can't tell me what the weather is going to be like in four months, or even next week?" Ball always elicited a knowing chuckle with this kind of commentary, which he delivered in hundreds of speeches across western Canada. No chamber of commerce, beef producers' association, or Probus Club of active retirees was too small to justify Ball's time. And the fact that Ball widely proclaimed himself to be "the first Canadian Ph.D. in climatology" seemed to give him carte blanche to confuse weather and climate—to dismiss out of hand the entire output of the best climate scientists working in the field today—and have his lunchtime crowds take him completely seriously.

Ball was the favorite front man for a Canadian Astroturf group called the Friends of Science—as Montgomery described them in the *Globe* article, "a coalition of oil-patch geologists, Tory [Canadian Conservative Party] insiders, anonymous donors and oil-industry PR professionals" from the Canadian oil capital of Calgary, Alberta. Here, taken from "Kyoto no no," one of Tim Ball's podcasts at fcpp.org, is his own entirely unwitting critique of the presumptuousness involved in calling your organization the "Friends of Science": "One of the things that angers me are these groups like Friends of Science. Now think of the arrogance of the title of that. Basically, what they are saying is that if you're not in our group, you're not a friend of science, or Friend of the Earth I should say. Sorry, the Friends of the Earth." Oops. According to Ball, it's okay to be an exclusive friend to science—even if your principal goal is to argue with the world's leading scientific experts—but it takes maddening arrogance to call yourself a friend of the Earth.

The Friends of Science motto is "Providing Insight into Climate Change," and on the Web site, friendsofscience.org, they announce their purpose this way: "Concerned about the abuse of science displayed in the politically inspired Kyoto protocol, we offer critical evidence that challenges the premises of Kyoto and present alternative causes of climate change."

Founded in 2002, Friends of Science attracted relatively little mainstream media attention and no serious independent scrutiny in its first couple of years. It made a series of YouTube videos such as "Climate Catastrophe Cancelled" and sponsored talks by Tim Ball and others. It solicited funds for the stated purpose of affecting Canadian elections.[1] But it wasn't until August 2006, when the freelance writer Charles Montgomery started nosing around on behalf of the *Globe and Mail*, that anyone really came to understand what the "Friends" were up to and who was paying their bills.

As Montgomery reported in "Meet Mr. Cool," you could argue that the Friends began as a legitimate grassroots organization:

> "We started out without a nickel, mostly retired geologists, geophysicists and retired businessmen, all old fogeys," says Albert Jacobs, a geologist and retired oil-explorations manager, proudly remembering the first meeting of the Friends of Science Society in the curling lounge of Calgary's Glencoe Club back in 2002.
>
> "We all had experience dealing with Kyoto, and we decided that a lot of it was based on science that was biased, incomplete and politicized."
>
> Mr. Jacobs says he suspects that the Kyoto Accord was devised as a tool by United Nations bureaucrats to push the world towards a world socialist government under the UN. "You know," he says, "to this day, there is no scientific proof that human-caused CO_2 is the main cause of global warming.

Given that Albert Jacobs was happy to rely on the likes of Tim Ball for scientific analysis—and giving him the benefit of the doubt—you could assume that he was sincere in his misguided position about the causes of global warming. But his ensuing fund-raising tactics demonstrated that he understood the implications of his actions. In the same *Globe* article Montgomery reported that the Friends were having difficulty garnering enough cash in the form of small donations to have the kind of impact that they wanted.

> "There was plenty of money for the anti-Kyoto cause in the oil patch, but the Friends dared not take money directly from energy companies. The optics, Mr. Jacobs admits, would have been terrible.
>
> This conundrum, he says, was solved by University of Calgary political scientist Barry Cooper, a well-known associate of Stephen Harper.
>
> As is his privilege as a faculty member, Prof. Cooper set up a fund at the university dubbed the Science Education Fund. Donors were encouraged to give to the fund through the Calgary Foundation, which administers charitable giving in the Calgary area, and has a policy of guarding donors' identities. The Science Education Fund in turn provides money for the Friends of Science, as well as Tim Ball's travel expenses, according to Mr. Jacobs.
>
> And who are the donors? No one will say.
>
> "[The money's] not exclusively from the oil and gas industry," says Prof. Cooper. "It's also from foundations and individuals. I can't tell you the names of those companies, or the foundations for that matter, or the individuals."

This was a sweet deal for the "Friends" and for their funders. The fledgling association got all the money it needed without having to account for its origins, and the oil and gas companies got tax receipts from the Calgary Foundation for donating to "educational" causes, even though the Friends' activities were overwhelmingly political rather than educational. But what had arguably started as a grassroots collection of retired oil-patch workers with a perhaps sincere belief in their position became an industry-funded political action group.

After the Montgomery story ran in the *Globe and Mail,* the Friends of Science became something of a political hot potato. The University of Calgary launched an internal audit that found Barry Cooper had sluiced hundreds of thousands of dollars through his "educational" accounts without meeting any of the university's standards for such actions. He had personally written checks as large as C$100,000 (to a Canadian arm of the public relations firm APCO Worldwide) and he had used some of the money to employ his wife and daughter, in contravention of University of Calgary policy. The University auditors also found that the uses to which the bulk of the money had been put "were not legitimate scientific research and education and were funded by anonymous donors to promote special interests."[2]

As punishment for all of this, Barry Cooper—fishing buddy of Canadian prime minister Stephen Harper—has paid no price whatsoever. He maintains his position at the university, and there is no record of a public reprimand or an order to make economic reparations. Cooper also still writes a weekly column, often on climate change politics, in the local "newspaper of record," the *Calgary Herald.*

Nevertheless, the Friends seem to be damaged goods generally, and though the Web site was still operating in early 2009 and the organization continues an annual luncheon, most of its activities have ground to a halt. But Tim Ball never missed a beat. Working with a former APCO public relations executive named Tom Harris, Ball reemerged in 2006 as the "chief science advisor" of a new organization called the Natural Resources Stewardship Project (NRSP), which, as a first order of business, promised to launch "A proactive *grassroots* campaign to counter the Kyoto Protocol and other greenhouse gas reduction schemes while promoting sensible climate change policy" (emphasis mine). Harris, who once had organized the public launch of the Friends of Science, stood in as executive director and, along with Ball, populated the list of other "science experts" with most of the scientists who had appeared on the Friends of Science honor roll.[3]

Tim Ball, who seemed during this period to get more work than ever, was famously reluctant to talk about who was paying his bills. For example, in August 2006 Harris flew Ball to Ottawa, where he briefed a committee in Canada's House of Commons. (This was a time when fiction writer Michael Crichton, author of the denial book *State of Fear,* was being asked to give climate change briefings to both then–U.S. president George W. Bush and the U.S. Senate Environment and Public Works Committee, then under the gavel of chair James Inhofe, a Republican from Oklahoma.) Before returning to his home on Canada's west coast, Ball stopped at the *Ottawa Citizen* for a meeting with that newspaper's editorial board. In a taped podcast of that meeting Ball says, "to my knowledge, I have never received a nickel from the oil and gas companies." Prodded about who was financing his cross-Canada speaking tour—picking up his expenses and paying him for his appearances—he says, "I made a point of not trying to find out who pays me."[4] Pressed further, Ball acknowledged that the High Park Group was paying his expenses in Ottawa. High Park, which had been Tom Harris's employer immediately before Harris launched the NRSP, states on its own Web site, highpark group.com: "Our client work is focused on natural resource and infrastructure sectors including power generation, energy transmission, transportation, and mining."

With a bit more investigation, it became obvious that High Park was more than a periodic funding source. Sleuthing about early in 2007, DeSmogBlog's Kevin Grandia discovered that two of the three directors on the board of the NRSP were senior High Park executives. Timothy Egan was the president of the High Park Advocacy Group and a retired lobbyist for the Canadian Gas Association and the Canadian Electricity Association. Julio Legos was the High Park Group's director of regulatory affairs, and his High Park biography said, "Julio's practice at HPG is focused on federal and provincial energy and environmental law and policy, particularly as they affect Canadian industry."[5, 6]

Creating the NRSP as an "arm's-length," "grassroots" organization enabled High Park to avoid identifying who was paying for the NRSP's public campaign against climate change regulations. The Canadian government restricts this kind of "grassroots" lobbying under the Lobbyists Registration Act, subject to an important loophole. The question and answer section under the "General Registration Requirements" of the act states:

> 4. What is "grass-roots" lobbying?
>
> Grass-roots lobbying is a communications technique that encourages individual members of the public or organizations to communicate directly with public office holders in an attempt to influence the decisions of government. Such efforts primarily rely on use of the media or advertising, and result in mass letter writing and facsimile campaigns, telephone calls to public office holders, and public demonstrations.

But in setting up the NRSP as a de facto subcontractor, High Park was able to claim an exception, which is also explained in the Q&A section of the registration requirements:

> 5. I am involved in organizing and directing a grass-roots lobbying campaign. Do I have to register?

> If you are a registered lobbyist, you must report grass-roots lobbying as a communications technique. If you are not engaged in any registerable lobbying activity, it is not necessary to register for the grass-roots lobbying campaign.[7]

So Tom Harris dropped off the lobbyist registry and moved his office across the hall (the NRSP mailing address was #2-263 Roncesvalles Avenue in Toronto; High Park's address was #4-263 Roncesvalles Avenue), after which he was able to carry out this "grassroots" campaign against energy industry regulation without incurring what the Lobbyist Act describes as the "obligation to provide accurate information to public office holders and to disclose the identity of the person or organization on whose behalf the representation is made and the purpose of the representation."

The federal lobbyists registry was created specifically so that politicians and members of the public can know who is paying to influence the political decision-making process. To that end the federal government's Lobbyist Code of Conduct says: "Lobbyists shall, when making a representation to a public office holder, disclose the identity of the person or organization on whose behalf the representation is made, as well as the reasons for the approach."

But Tom Harris is not technically a lobbyist, and Timothy Egan and Julio Legos may well have been "volunteering" their time as directors of the NRSP.

So during this period Egan was a registered lobbyist for the Canadian Gas Association, which is part of an energy industry coalition that includes the Canadian Nuclear Association, the Canadian Association of Oil Well Drilling Contractors, the Canadian Energy Alliance, the Propane Gas Association of Canada, the Petroleum Services Association of Canada, the Canadian Association of Petroleum Producers, the Canadian

Petroleum Products Institute, the Canadian Energy Pipeline Association, and the Coal Association of Canada, as well as some conservation and alternative-energy interests such as the Canadian Energy Efficiency Alliance, the Canadian Wind Energy Association, and Hydrogen and Fuel Cells Canada. But regardless of that status, and irrespective of the NRSP's stated purpose to block government action on climate change, the nearly arm's-length relationship between High Park and the NRSP meant that the Canadian public had no right to ask who was paying the bills for the NRSP campaign.

Harris continued to run the NRSP through all of 2007, even after Canada's largest newspaper, the *Toronto Star,* ran a feature ("Who's Still Cool on Global Warming?") on January 28, 2007, exposing his energy-industry connections. But in March 2008 he popped up as the new executive director of the International Climate Science Coalition, which operates from the same IP address and with most of the same "experts" (hello, Tim Ball and Albert Jacobs) as the New Zealand Climate Science Coalition and the Australian Climate Science Coalition. The International Climate Science Coalition Web site, climatescienceinternational .org, states that the group "is an international association of scientists, economists and energy and policy experts working to promote better public understanding of climate change science and policy worldwide. ICSC is committed to providing a highly credible alternative to the UN's Intergovernmental Panel on Climate Change (IPCC) thereby fostering a more rational, open discussion about climate issues."

Although Harris doesn't tend to invoke the word "grassroots" quite so often at the International Climate Science Coalition, he remains committed to the old tactics. At a strategy session at the "2008 International Conference on Climate Change," a gathering of climate deniers organized by the Heartland Institute, Harris said this (emphasis added):

> We need regular high-impact media coverage of the findings of leading scientists—not just one or two publications, but we need to have hundreds all over the world. We need to have a high degree of information sharing and cooperation between groups, so that when [the well-known New Zealand climate change denier] Vincent Gray, for example, has an article published in New Zealand, we can take the same piece and we can submit it to newspapers all over North America and Europe.
>
> Then we have a nicely well-coordinated response, where letters to the editor and phone calls are made. "Congratulations on publishing that article!" You know, it's interesting because I've had many of my articles opposed so strongly, by environmentalists through phone calls and letters to the editor, that they just simply dry up, they just won't publish us again. So this does have feedback, I mean, these are people that run these newspapers, and they're scared, and impressed, and encouraged, depending on the feedback they get.
>
> We have to have *grassroots organizations* doing exactly that kind of thing: coordinated local activism.
>
> And finally, as I said, we need unbiased polling and good press coverage.[8]

Nowhere in any of this material does Harris say that we need good science. The kind of articles that Vincent Gray is writing these days don't turn up in the reputable pages of *Science* or *Nature*. They appear on the opinion pages of middle-market newspapers, whence they are photocopied and submitted to other middle-market newspapers, as Harris says, "all over North America and Europe." And if one of those newspapers actually runs the article again, Harris has his "grassroots" legions follow up with flattering phone calls and congratulatory letters.

We are left, then, with a trail of misdirection. It began with an organization (the Friends of Science) whose principals admitted they were trying to cover up their funding sources. It transformed into a second organization that was clumsy enough to allow its sources to be exposed. And that transformed into a new and more international organization that is not directly subject to Canadian or American laws demanding financial disclosure. Then again, transparency has never been the order of the day. After the University of Calgary audit showing the adventures of Professor Barry Cooper—the transparent attempt to hide the source of oil-and-gas industry funding—no one paid a price. No taxes were recaptured from people who had received tax receipts to which they were not legally entitled. No wrists were slapped. The University changed some of its internal processes, but the government of Canada dropped its investigations without prosecuting or imposing tax penalties.

The Friends of Science also dodged a bullet on their campaign activities. Having solicited money and targeted swing ridings for a radio advertising campaign during the 2006 Canadian federal election, after stating publicly that its purpose was "to have a major impact on the next election,"[9] the Friends of Science were absolved of responsibility by Elections Canada. Without explanation, and certainly with no mention of the relationship between Prime Minister Stephen Harper and Barry Cooper—the federal regulatory agency announced that it was dropping its investigation and planned no further action, even though on the face of it, the Friends' activities were apparently designed to affect a federal election in a way that is specifically proscribed in law.[10]

Thus we have a series of "grassroots" organizations whose actual roots appear to track directly to the fossil fuel industry. We also have a continuing campaign, first outlined so clearly in the Western Fuels Association, TASSC, and API strategy documents, to sow doubt and confusion—to raise questions about science in the public mind without ever actually embracing or referencing the work of leading scientists in the field. Once again, if doubt reigns in the minds of the public, it appears to be anything but accidental.

[*six*]

MANGLING THE LANGUAGE

Making doubt reliable and science unbelievable

Now, it is clear that the decline of a language must ultimately have political and economic causes: it is not due simply to the bad influence of this or that individual writer. But an effect can become a cause, reinforcing the original cause and producing the same effect in an intensified form, and so on indefinitely. A man may take to drink because he feels himself to be a failure, and then fail all the more completely because he drinks. It is rather the same thing that is happening to the English language. It becomes ugly and inaccurate because our thoughts are foolish, but the slovenliness of our language makes it easier for us to have foolish thoughts.

GEORGE ORWELL, "POLITICS AND THE ENGLISH LANGUAGE"

There must have been people before George Orwell who recognized and deplored the use of language to hide meaning—to divert, deceive, or confuse rather than to illuminate. But when Orwell wrote the words above in 1946, he identified a tactic that offended both his thirst for clarity and his lust for justice. Orwell complained about the increasing popularity of the euphemism, which the second edition of the *Oxford English Reference Dictionary* defines as "a mild or vague expression substituted for one thought to be too harsh or direct." In the same essay, Orwell offers these as examples:

> Defenseless villages are bombarded from the air, the inhabitants driven out into the countryside, the cattle machine-gunned, the huts set on fire with incendiary bullets: this is called *pacification*. Millions of peasants are robbed of their farms and sent trudging along the roads with no more than they can carry: this is called *transfer of population* or *rectification of frontiers*. People are imprisoned for years without trial, or shot in the back of the neck or sent to die of scurvy in Arctic lumber camps: this is called *elimination of unreliable elements*. Such phraseology is needed if one wants to name things without calling up mental pictures of them.

In this age of Peacekeeper missiles and collateral damage, Orwell likely would be complaining more bitterly than ever. And seeing the emerging corporate appetite for newspeak words like "downsizing" and "outsourcing," he wouldn't be able to help noting that the manipulation of language had spread from the political to the corporate sphere. He might even be angered that the people who practice this manipulation today have gained some of their insight and honed their skills from reading the works that Orwell himself wrote in good faith, in an effort to root out, rather than encourage, campaigns of misinformation.

One of Orwell's big fans in America today is the Republican pollster and spin doctor (a descriptor I use advisedly) Frank Luntz. In a January 9, 2007, interview with Terry Gross on the National Public Radio show *Fresh Air*, Luntz said that "the average American assumes that being Orwellian means to mislead," almost assuredly a reference to Orwell's depiction in novels such as *Nineteen Eighty-Four* of the intentional and misleading rewriting of history. But pointing to what he referred to as "Orwell's essay on language," likely "Politics and the English Language," Luntz said that "to be 'Orwellian' is to speak with absolute clarity, to be succinct, to explain what the event is, to talk about what triggers something happening... and to do so without any pejorative whatsoever."

That honors Orwell's memory more than it reflects the accepted meaning of the word. Wikipedia, for example, says as of April 29, 2009, that "Orwellian describes the situation, idea, or societal condition that George Orwell identified as being destructive to the welfare of a free society. It connotes an attitude and a policy of control by propaganda, misinformation, denial of truth, and manipulation of the past." And while Luntz says that he would prefer "Orwellian" to mean something else, the way he uses the language often invokes the darker and more widely accepted definition.

Consider, for example, something else Luntz said in the same interview. He rose to defend the use of the term "energy exploration" to refer to oil drilling in the Arctic National Wildlife Refuge. He then gave a brief description of what he imagines "energy exploration" looks like. Noting that "98 percent of exploration happens underneath the ground," Luntz pointed out that when this "exploration" is over, if no oil is discovered, the only remaining evidence is a couple of feet of pipe sticking out of the ground. Thus, he said, using the term "exploring" is "more precise, more 21st century. It's cleaner, more careful." To affirm this view, Luntz said that he gathered focus groups and showed them photographs of drilling operations. He asked participants if those photos suggested "energy exploration" or "oil drilling," and 88 percent of them said "energy exploration." He wrapped up his point by saying, "Therefore, I'd argue that it is a more appropriate way to communicate... If the public says after looking at the pictures, 'that doesn't look like my definition of drilling; it looks like my definition of exploring,' then don't you think we should be calling it what people see it to be, rather than adding a political aspect to it all?"

Luntz's ability to nudge the language for his own purpose and on occasion to get people to agree with his manipulations calls forth Orwell's comment that "our thoughts are foolish, but the slovenliness of our language makes it easier for us to have foolish thoughts." It also makes you wonder what Luntz was showing people in those photos.

Luntz also said in an interview on the C-SPAN book-review show *After Words* that "deep-sea energy exploration" is a more precise term than "oil drilling" because "drilling suggests that oil is pouring into the ocean. In Katrina, not a single drop of oil spilled in the Gulf of Mexico from the rigs themselves. That's why deep-sea exploration is a more appropriate term."[1]

Again, that's an interesting bit of spin that would be compelling if it was true. Det Norske Veritas, the firm that conducted the final appraisal of oil-related damage from hurricanes Katrina and Rita, showed that there were 124 offshore spills that contaminated the ocean with 743,700 gallons of oil, of which 554,400 gallons were crude oil and condensate from platforms, rigs, and pipelines and 189,000 gallons were refined products from platforms and rigs.[2]

Luntz's most famous linguistic manipulation, however, is the *Straight Talk* memo that he wrote in 2002 as part of a major briefing package preparing the Bush White House for the coming elections. Titled "The Environment: A Cleaner, Safer, Healthier America," the paper displays a mastery of Orwellian language, both by Luntz's preferred definition (the language is clear and precise) and by the more conventional sense of Orwellianism: it is shameless but incredibly creative in the way it counsels Republican candidates to distract or misdirect public attention while talking about the environment.

Luntz begins this overview by saying, "The environment is probably the single issue on which Republicans in general—and President Bush in particular—are most vulnerable." He then proceeds through sixteen pages of advice that suggests that the problem is not the Republican record of dismantling environmental protections and fighting international treaties, but rather the language that Republicans have used to describe these actions.

For example, this is Luntz's opening comment on climate change (I have retained all of the underlined, italicized, and boldfaced emphasis that Luntz included in the original):

> 1. ***The scientific debate remains open.*** Voters believe that there is ***no consensus*** about global warming within the scientific community. Should the public come to believe that the scientific issues are settled, their views about global warming will change accordingly. Therefore, ***you need to continue to make the lack of scientific certainty a primary issue in the debate,*** and defer to scientists and other experts in the field.

In 2002 the statements in this paragraph were all, strictly speaking, correct: a majority of voters really did believe there was no consensus about climate change in the scientific community. Of course, there was overwhelming scientific consensus, but Luntz was not commenting on reality, he was giving the Republicans guidance about perception. And the perception created a clear opportunity for the Republicans to continue using this public confusion to their advantage. Luntz even pursued this line on the next page, saying,

> ***The scientific debate is closing*** [against us] ***but not yet closed. There is still a window of opportunity to challenge the science.*** Americans believe that all the strange weather that was associated with El Niño had something to do with global warming, and there is little you can do to convince them otherwise. However, only a handful of people believes the science of global warming is a closed question. Most Americans want more information so that they can make an informed decision. It is our job to provide that information.

At no point does Luntz urge that the information should be correct, up-to-date, or broadly supported in the scientific community. Rather, he says, ***"You need to be even more active in recruiting experts who are sympathetic to your view, and much more active in making them part of your message."***

The memo is also filled with little breakout boxes enshrining what Luntz calls "Language That Works" (a precursor, perhaps, to the book that he was working on, *Words That Work*). One such box suggests that Republican candidates should be saying things like this:

> ***Unnecessary environmental regulations hurt moms and dads, grandmas and grandpas. They hurt citizens on fixed incomes. They take an enormous swipe at miners, loggers, truckers, farmers—anyone who has any work in energy intensive professions. They mean less income for families struggling to survive and educate their children.***

"Moms and dads, grandmas and grandpas." The cynicism in this is surely, darkly Orwellian. But at every turn Luntz shows that he has based his advice on careful research. He has tested his messages in focus groups and through opinion polls, any one of which would have strained the resources of the environmental groups that were at the time trying to balance this presidential information source with accurate countercommentary. Drilling (or maybe we should say "exploring") further into the language, Luntz says:

> 1. ***"Climate change" is less frightening than "global warming."*** As one focus group participant noted, climate change "sounds like you're going from Pittsburgh to Fort Lauderdale." While global warming has catastrophic connotations attached to it, climate change suggests a more controllable and less emotional challenge.

2. ***We should be "conservationists," not "preservationists" or "environmentalists."*** The term "conservationist" has far more positive connotations than either of the other two terms. It conveys a moderate, reasoned, common sense position between replenishing the earth's natural resources and the human need to make use of those resources.

Thus we have moderate, reasoned, and commonsense language to describe immoderate, unreasonable, and insensible government policy. A year after President George W. Bush asked the National Academy of Sciences to report on the risks of climate change—which is to say, a year after the academy assured the president that the risk was real and increasingly severe—one of the cleverest linguistic manipulators in the country provided talking points that would enable the president to convince Americans to adopt a "commonsense" position based on ignorance and self-interest.

Perhaps the defining statement in Luntz's commentary—and the one that links it most closely to its tobacco-era roots—is this: "The most important principle in any discussion of global warming is your commitment to ***sound science***." I would have put "sound science" in boldfaced italics if Luntz hadn't done so already. He is not referring, in the clear and precise way that Orwell himself might have preferred, to science. Not peer-reviewed science. Not reliable science that is supported by the overwhelming majority of climate experts in the United States and around the world. It's something else. It is "sound science," and "sound" is obviously a well-tested adjective. Orwell might have shuddered at the usage.

Mind you, Orwell probably would have shuddered just as violently about the corresponding standard of concision and clarity in the scientific writing. The scientists in the IPCC are appropriately, obsessively precise, but sometimes that doesn't make their work any easier to understand. For example, the

IPCC's *Climate Change 2007: Synthesis Report*, available at www.ipcc.ch, includes a five-hundred-word essay on the "Treatment of Uncertainty." Here's an example of their version of absolute clarity:

> Where uncertainty is assessed qualitatively, it is characterised by providing a relative sense of the amount and quality of evidence (that is, information from theory, observations or models indicating whether a belief or proposition is true or valid) and the degree of agreement (that is, the level of concurrence in the literature on a particular finding). This approach is used by WG III through a series of self-explanatory terms such as: high agreement, much evidence; high agreement, medium evidence; medium agreement, medium evidence; etc."

You might argue that if terms are indeed self-explanatory, they shouldn't need a five-hundred-word essay to explain them, but such is the gap between the scientist and the public relations professional. In the same section, the IPCC does a slightly clearer job of defining what it means by certainty and uncertainty in another paragraph:

> Where uncertainty in specific outcomes is assessed using expert judgment and statistical analysis of a body of evidence (e.g. observations or model results), then the following likelihood ranges are used to express the assessed probability of occurrence: virtually certain >99%; extremely likely >95%; very likely >90%; likely >66%; more likely than not >50%; about as likely as not 33% to 66%; unlikely <33%; very unlikely <10%; extremely unlikely <5%; exceptionally unlikely <1%.

Such is the caution of the professional scientist that the IPCC now speaks about human-induced global warming as "very likely," leaving a margin of doubt that allows people like Frank Luntz to point helpfully to the remaining element of doubt. Interestingly, however, no one on the Luntz team ever mentioned the obvious: that if scientists told you there was a 90 percent likelihood that your plane would crash, you would assuredly forego the trip. But when the conveyance of choice is planet Earth, as Luntz said, "You need to continue to make the lack of scientific certainty a primary issue in the debate." Again, his concern was perception and political advantage, not consequences.

In fairness, at the time Luntz wrote his *Straight Talk* memo, he was still relying on the results of the IPCC's previous assessment report, from 2001, which had stated that the science could only pronounce that the human influence in changing the world climate was "likely." That in part was what allowed Luntz to proclaim that "the scientific debate remains open."

Luntz's analysis was impressive from a technical standpoint. It is very sharp, very clever, and it's clear from President Bush's re-election in 2004 that he and his party dealt well with this, one of their weakest issues. It's also worth noting that this kind of work is relatively common in public relations. Many corporations, trade associations, and politicians employ pollsters and analysts to run focus groups and test words, phrases, and key messages. Until you do so, it is sometimes very difficult to know whether your target audience is hearing and understanding what you *think* you are saying. But there is a presumption that what you are saying is either objectively true or fairly represents a legitimate opinion.

That presumption may have been too optimistic when the Western Fuels Association was working the ICE file, a process that also involved a significant front-end investment in testing words and messages. As Naomi Oreskes writes in a chapter titled "My Facts Are Better than Your Facts: Spreading Good News about Global Warming," the Western Fuels Association even tested the name of their new Astroturf group. They thought ICE was a good acronym, but couldn't decide whether to call their organization the "Information Council for the Environment," "Informed Citizens for the Environment," "Intelligent Concern for the Environment," or "Informed Choices for the Environment." Oreskes writes, "The focus groups indicated that American citizens trusted scientists more than politicians or political activists—and industry spokesmen least—so Western Fuels settled on *Information Council for the Environment,* because it positioned ICE as a 'technical' source rather than an industry group."[3]

As I mentioned in Chapter 4, some of the Western Fuels Association's messages were aimed primarily at making the notion of climate change sound silly ("Some say the Earth is warming. Some also said the Earth was flat."), but they also tested others that were "fact" specific. For example, they tried out, "If the Earth is getting warmer, why is Minneapolis getting colder?" and they tested, "If the Earth is getting warmer, why is the frost line moving south?"

Again, the first test for these messages should have been whether they were true. Minneapolis was *not* getting colder. The frost line was *not* moving south. And all but the scientists whom the Western Fuels Association was paying to say otherwise seem to have agreed, even then, that the Earth was getting warmer and that people were to blame. But the Western Fuels Association was not testing for facts. It was testing the tolerance and responsiveness of its target audience. It was also clear in its agenda, which it summed up in three points:

1. To demonstrate that a "consumer-based awareness program can positively change the opinions of a selected population regarding the validity of global warming";

2. to "begin to develop a message and strategy for shaping public opinion on a national scale"; and
3. to "lay the groundwork for a unified national electric industry voice on global warming."

The target audience in question was nested in four cities: Chattanooga, Tennessee; Champaign, Illinois; Flagstaff, Arizona; and Fargo, North Dakota. These were chosen because they got most of their electricity from coal, they each were home to a member of the U.S. House of Representatives Energy and Commerce or Ways and Means Committees, and they had low media costs, which meant that it was going to be cheap to test the national campaign.

These are all wonderful details. They show a real degree of thoughtfulness, even professionalism, on the part of the people who designed the program. As with the Luntz analysis, you can see the intelligence, even tactical brilliance, that went into the campaign. What you cannot see is any evidence that anyone, at any time, asked whether what they were doing was right—whether, for example, the messages they were testing could have been incorrect and ultimately harmful.

Luntz, in the previously quoted radio conversation with the National Public Radio's Terry Gross, continued throughout the interview to defend his use of language—even to suggest that what he was doing was a good and necessary thing. He said, "Corporations, trade associations, and politicians have a responsibility to communicate in a way that makes it most likely that the public will support where they stand."

Really? Don't corporations, trade associations, and politicians have a responsibility to communicate in a way that is fair, honest, and in the public interest? Do we assume that because it proved effective in the Western Fuels Association focus group testing, presenting incorrect information to the public about the actual details of climate change is a *responsible* option?

Luntz himself recanted on the climate file in 2006. He told the BBC: "It's now 2006. I think most people would conclude that there is global warming taking place and that the behavior of humans is affecting the climate." When the BBC reporter said, "But the administration has continued taking your advice; they're still questioning the science," Luntz responded, "That's up to the administration. I'm not the administration. What they want to do is their business. It has nothing to do with what I write. It has nothing to do with what I believe." That seems tantamount to saying, "I just work here" or, in the dark Orwellian version, "I was just following orders."

I might have hoped for something more. When the President of the United States asks you to prepare a briefing note about communicating on the environment, I would argue that you have a responsibility to answer honestly—to inform yourself about the issue and to ensure that you are working in the service of your political master *and* in the service of the people who put your master into office. The "just following orders" defense was never a very good one, but it sounded more convincing coming from people who added an apology—even after the fact. Luntz apparently believes that he owes no apology, that he has done nothing wrong. That might give you fair warning the next time you hear that someone is buying his advice.

[*seven*]

THINK TANK TACTICS

Moving public policy into private hands

In November 2006 a senior academic and climate scientist at a major U.S. university passed on a letter that had been distributed with the signature of Kenneth Green, a resident scholar at the American Enterprise Institute. Green was offering cash to scientists who would agree to write a critique of the anticipated Fourth Assessment Report of the IPCC. The thrust of the letter is evident from this paragraph:

> The purpose of this project is to highlight the strengths and weaknesses of the IPCC process, especially as it bears on potential policy responses to climate change. As with any large-scale "consensus" process, the IPCC is susceptible to self-selection bias in its personnel, resistant to reasonable criticism and dissent and prone to summary conclusions that are poorly supported by the analytical work of the complete Working Group reports.[1]

The American Enterprise Institute is one of a battery of think tanks that have in the past decade or more received significant funding from ExxonMobil and have, perhaps coincidentally, taken a highly public position challenging the consensus that human activity is causing climate change. In this instance Green was offering any willing scientist US$10,000 plus expenses to write a critique. Green also suggested that American Enterprise Institute would be happy to host a "series of small conferences and seminars in Washington and elsewhere" and indicated a further willingness to pay additional expenses and honoraria to any scientist willing to participate. Long before the Fourth Assessment Report was released, the institute had apparently decided that it was not going to like the result and was eager to find scientists who would help criticize the findings, whatever they might be.

By way of background, the IPCC was founded in 1988 by the United Nations Environment Programme in cooperation with the World Meteorological Organization, with a stated goal of assessing scientific information relevant to the following concerns:

1. Human-induced climate change;
2. The impacts of human-induced climate change; and
3. Options for adaptation and mitigation.

As the Nobel committee noted when it gave the IPCC (along with former U.S. vice president Al Gore) the Nobel Peace Prize in 2007, the panel is one of the most impressive collaborations in global scientific history—basically a blue-ribbon review panel comprising the very best talent from every corner of the world. At intervals of three to four years these scientists, almost all of whom are currently engaged in leading-edge climate research, pull together the most reliable peer-reviewed scientific information and, in separate working groups dealing with different aspects of the issue, hammer out reports that are written to the most exacting standards. These reports are then condensed into a *Summary for Policymakers,* which is subject to a review by participating governments as well as by participating scientists.

The latter process is largely to be blamed for opening the IPCC up to legitimate accusations of political interference. The consensus format tends to give the greatest influence to the most resistant parties. If there is the tiniest grain of doubt in any specific piece of science, it is likely to be dismissed, either in the last scientific review or in the first political one. When you consider that among the reviewers you have the governments of oil-producing giants such as Russia, Saudi Arabia, and Argentina, you can imagine a degree of foot-dragging. Add to that the fears of India and China that they will be prohibited from lifting their nations out of poverty, and, perhaps worse, the intractability over the past eight years of the Bush administration, and you have a review process that was indeed highly politicized and that strained the scientists' ability to put a sensible and accurate document before the people of the world.

In that light, it is—what's that phrase again?—Orwellian in the extreme to suggest that the IPCC was biased toward overstating the risks of selling and burning oil, coal, and natural gas. But that was the unavoidable inference that an impartial reader might draw from Ken Green and his Exxon-sponsored think tank compatriots. After receiving a copy of Green's letter of solicitation, the researchers at the DeSmogBlog started watching for evidence that any scientists had taken up the challenge.

Round two began at the end of January 2007, barely a week before the scheduled release of the first, and potentially most controversial, section of the IPCC's Fourth Assessment Report. A Canadian think tank, the Fraser Institute, announced that it was about to release an *Independent Summary for Policy Makers* (*ISPM*). This seemed like more than a passing coincidence. Ken Green's last job before moving to the American Enterprise Institute was Director for the Centre of Studies of Risk and Regulation at the Fraser Institute. And the Fraser Institute is also listed on Greenpeace's ExxonSecrets Web site as a recipient of direct funding from ExxonMobil and other energy interests.

Although the Fraser Institute had promised to release its summary on February 5, 2007, Kevin Grandia at DeSmogBlog obtained a copy on January 31, 2007. It had been circulated for its own informal "peer review," and one of the potential reviewers was sufficiently concerned about the content to pass it along for early release. We obliged, posting it on DeSmogBlog for a somewhat wider public review.

The report itself was unsurprising. Although it stated that "the ISPM was prepared by experts who are fully qualified and experienced in their fields, but who are not themselves IPCC chapter authors," the actual credits showed that the project coordinator was an economist, Ross McKitrick (a senior fellow at the Fraser Institute), and the lead author was the Weather Channel's retired chief meteorologist, Joseph D'Aleo, a man who had had been working on a Ph.D. at New York University when he chose to leave academia.

The *Independent Summary for Policy Makers* was long (more than fifty pages), convoluted, and obsessed with uncertainty. Its most enthusiastic argument addressed the so-called Mann hockey stick graph, which had appeared in an earlier IPCC document but was not part of the *Fourth Assessment Report* purportedly under review. The summary's overall conclusion, which had been telegraphed by Green's original letter of solicitation, was this: "There is no evidence provided by the IPCC in its Fourth Assessment Report that the uncertainty can be formally resolved from first principles, statistical hypothesis testing or modeling exercises. Consequently, there will remain an unavoidable element of uncertainty as to the extent that humans are contributing to future climate change, and indeed whether or not such change is a good or bad thing."

Notwithstanding that actual climate scientists—the best in the world—had judged there to be a more than 90 percent chance that our spacecraft is headed for trouble, the American Enterprise and Fraser institutes, and the "scientists" who were prepared to take their money, recommended that we embrace that less than 10 percent "unavoidable element of uncertainty" and continue apace.

After we released the report five days early, we also contacted friends in London, England, where Ken Green and company had scheduled a press conference for the following Monday. (This itself was curious. The IPCC report was released in Paris. The Fraser Institute is based in Vancouver, B.C., and the American Enterprise Institute is in Washington, D.C. We wondered if they thought the London media wouldn't have heard about the origin of the report.) When the event finally occurred, our London contacts distributed background information on the participants, and the few reporters who turned up seemed unmoved by the Fraser Institute's analysis. If Exxon got any news coverage for this particular investment, we couldn't find it.

This seemed like an excellent example, however, of some of the activities of a large group of "think tanks" that have come under public scrutiny for accepting funding from major industrial sources such as ExxonMobil and then challenging the science of climate change. I put the term "think tank" in quotes because it is so difficult to even define these organizations accurately, much less understand what they do. The original think tanks were founded as policy-development hothouses. They were organizations such as the Brookings Institution (founded 1916) in the U.S. or the C.D. Howe Institute (founded 1958) in

Canada, where participants undertook research on issues of public interest and then served up the results for political and sometimes public consideration. The actual term "think tank" is thought to have emerged in the 1950s in reference to military intelligence organizations like the RAND Corporation (RAND is an acronym for "Research and Development"), which were established to try to advise government on how to keep the Cold War from going nuclear.

By the 1970s and '80s think tanks were popping up like spring crocuses and seemed increasingly dedicated not to conceiving new policies, but rather to advocating for the kinds of policies that would advance the interests of their funders. We wound up with organizations like the Heartland Institute (founded in the Orwellian year 1984), which used its funding from Philip Morris not to consider whether smoking was a good thing, but to convince the public and, especially, the nation's legislators that regulating against smoking was a bad thing. Even today (at least as of April 2009), Heartland maintains a pro-tobacco "Smokers' Lounge" on its Web site, though it now refuses to acknowledge where the funding for this feature comes from. In short, for at least some of these institutions the emphasis has shifted from "think" to "tank." What were once centers of academic excellence have become heavy-duty weapons in the battle for public opinion and political support.

This was fairly clear in the pro-tobacco performance of the Heartland Institute, but it was more difficult to establish on the climate change file. Greenpeace research director Kert Davies said in an interview with Richard Littlemore in February 2009 that his organization was becoming increasingly frustrated by the non-response from mainstream media to what Greenpeace felt was an obvious, think tank–driven manipulation of the public conversation. Greenpeace had found and publicized the American Petroleum Institute's "Global Climate Science

Communications Plan," which showed that four of the most prominent climate change–denying think tanks (the Heartland Institute, George C. Marshall Institute, American Legislative Exchange Council, and the Frontiers of Freedom Institute) were involved in conceiving that plan. But whenever Davies or his colleagues tried to point out the connection, reporters shrugged it off. No corpse, no bloodstained assassin, no story.

So Greenpeace researchers did what any conspiracy theorist (or smart police detective) does in this situation: they followed the money. A group called the Clearinghouse on Environmental Advocacy and Research had started to build a database in the 1990s that included records of major industry contributions to the wise use movement, a property-rights coalition that had been opposing environmental regulations ranging from wetland protection to the U.S. Endangered Species Act. Greenpeace picked up the group's material and research techniques and started composing a specific database of contributions that ExxonMobil was making to think tanks—and especially to think tanks participating in the campaign to promote uncertainty about climate change.

Davies said Greenpeace picked Exxon for two reasons. First, as the biggest and wealthiest corporation in the world, it had the capacity to exercise an immense influence. And second, it was the only major oil company in the world that had not by the turn of the century acknowledged that climate change was a problem and that the burning of fossil fuels was the principal cause.

The Greenpeace team set up a Web site called ExxonSecrets .org. Based on tax information and on Exxon's own annual corporate giving report, ExxonSecrets laid out which think tanks were receiving money and how much. Through creative use of "mind-mapping" display technology, the site also drew connections among the institutions and "experts" who were making themselves famous (and increasing their incomes) by claiming that the science of climate change was still in doubt. For example, the scientist Dr. S. Fred Singer, a hardworking climate change denier who has done no obvious scientific work in the field for years, was shown as president of his own think tank, the Science & Environmental Policy Project; editorial advisory board member for the Cato Institute; advisory board member for the American Council on Science and Health; adjunct scholar for the National Center for Policy Analysis; research fellow for the Independent Institute; distinguished research professor at the Institute for Humane Studies, George Mason University; former adjunct fellow at the Frontiers of Freedom Institute; former fellow at the Hoover Institution; former fellow at the Heritage Foundation; former fellow at TASSC; and editor of the newsletter *Global Climate Change.* ExxonSecrets also showed that all of these organizations receive money directly or indirectly from Exxon.

ExxonSecrets established that in the ten years after the creation of the Kyoto Protocol, Exxon invested more than US$20 million in think tanks that dedicated a large amount of effort to questioning whether climate change was sound science.

While Greenpeace was nailing down the dollars, three academics were conducting a peer-reviewed study of actual think tank output from the early 1970s forward. Peter Jacques and Mark Freeman of the Department of Political Science at the University of Central Florida in Orlando and Riley Dunlap of the Department of Sociology at Oklahoma State University–Stillwater looked at the rise in publications that express all kinds of "environmental skepticism." Their paper, titled "The Organization of Denial: Conservative Think Tanks and Environmental Scepticism" and published in June 2008 in the journal *Environmental Politics,* searched all the available English-language books published between 1972 and 2005 that denied the seriousness of environmental problems. They found 141 such books denying or downplaying the seriousness of issues including climate change; stratospheric ozone depletion; biodiversity loss; resource shortages; chemical and other pollutants in the air, water, or soil; threats to human health of trace chemical exposure; and the potential risks of genetic manipulation. Of those 141 books, 130 (92.2 percent) were published by conservative think tanks, written by authors affiliated with those think tanks, or both.

Jacques, Dunlap, and Freeman concluded that "environmental skeptics are not, as they portray themselves, independent and objective analysts. Rather, they are predominantly agents of conservative think tanks, and their success in promoting skepticism about environmental problems stems from their affiliation with these politically powerful institutions."

As for the think tanks themselves, Jacques and company also argued that the batch of environmentally skeptical organizations were in a class by themselves:

> Unlike traditional think tanks that aimed to provide reasonably "objective" policy analyses, CTTs [conservative think tanks] are "advocacy" organisations that unabashedly promote conservative goals. Launched in the 1970s in reaction to social activism and an expanding federal government, CTTs were an institutional answer from American business leaders who during this time "voiced fears of 'creeping socialism.'" The strategy was to create an activist counter-intelligentsia to conduct an effective "war of ideas" against proponents of government programmes designed to ameliorate presumed social problems such as poverty.

Thus, again the think tanks were not primarily engaged in research, but were acting as strategists and lobbyists in the "war of ideas." The think tanks on the Jacques list also did more than publish books or other materials intended to influence public opinion or policy. In the most dramatic instance, the Competitive Enterprise Institute (CEI) twice sued the U.S. government in an effort to block the release of the *National Assessment of Climate Change*, a comprehensive report on likely climate change effects and complications that was commissioned under the Clinton Administration and slipped into a drawer during the Bush years.

Jacques and company also attempted a random assessment of the material that the think tanks were promoting on their Web sites, aside from what they were pushing into print. The academics chose fifty conservative think tanks, gleaned from a list posted by the Heritage Foundation, which was established in 1973 and has stood as the granddaddy of conservative think tanks ever since. Then they checked the Web site content for material that promoted skepticism on environmental issues. Of the fifty identified, forty-five (90 percent) espoused environmentally skeptical policies. And of the forty-three skeptical think tanks that Jacques lists specifically in his paper, twenty-eight show up on the ExxonSecrets list as having accepted money from ExxonMobil in the past ten years.

Through the early years of the new millennium, there is little doubt that the CEI was the leading vehicle in the anti–climate-science think tank battalion. CEI was the biggest recipient of Exxon funding, taking more than US$2 million of the US$20-plus million that Exxon spent on denial beginning in 1998. A leaked coal-industry memo also reported that GM and Ford were frequent funders of CEI's climate change activities, although Ford specifically denied that it had funded CEI's infamous series of TV commercials lauding the burning of fossil fuels and celebrating the production of carbon dioxide.[2]

These commercials, which as of February 2009 were still available for viewing on the CEI Web site, were beautifully and expensively produced. They were also so over-the-top that they probably did CEI more damage than good. Featuring bubble-blowing beauties and lushly attractive photos of the sun rising over a distant oil refinery, the ads challenged evidence that carbon dioxide is creating any kind of problem in the world. Far from being a pollutant, the ads said, carbon dioxide is "essential to life." With orchestral chords building in the soundtrack, the announcer intoned, "We breathe it out; plants breathe it in." In the final scene a little girl blows the seeds off a single, perfect dandelion while the announcer says, "Carbon dioxide. They call it pollution. We call it life."

What's missing here is a sense of scope. Iron is an essential vitamin, but that would not make a diet of nails any more palatable.

There is no specific evidence of what pushed the Royal Society of London to object to CEI's nonsense, but in September of 2006 news broke that this, one of the most august scientific bodies in the world, had called Exxon onto the carpet and demanded that it stop funding the denial machine and stop quibbling about the science. In a letter to Exxon that was leaked to the press and published in the U.K. *Guardian,* Bob Ward of the Royal Society said:

> It is now more crucial than ever that we have a debate which is properly informed by the science. For people to be still producing information that misleads people about climate change is unhelpful. The next IPCC report should give people the final push that they need to take action and we can't have people trying to undermine it ... At our meeting in July ... you indicated that ExxonMobil would not be providing any further funding to these organizations. I would be grateful if you could let me know when ExxonMobil plans to carry out this pledge.

In a response reported in the *Guardian* on September 20, 2006, under the headline "Royal Society Tells Exxon: Stop Funding Climate Change Denial," an unnamed Exxon executive replied, "We can confirm that recently we received a letter from the Royal Society on the topic of climate change. Amongst other topics our Tomorrow's Energy and Corporate Citizenship reports explain our views openly and honestly on climate change. We would refute any suggestion that our reports are inaccurate or misleading." But by January 2007 Exxon was changing its tune on climate change and on funding the think tanks that had earned the Royal Society's contempt. "We know enough now—or society knows enough now—that the risk is serious and action should be taken," Exxon Vice President for Public Affairs Kenneth Cohen told the *Wall Street Journal*. In particular, Cohen announced that Exxon had cut off funding to the CEI.

That was all to the good, except that when Greenpeace checked the record later the same year, it found that Exxon during 2006 had still divided more than US$2 million among forty-one other think tanks active in the denial movement. And in May 2008, when Exxon released its *2007 Corporate Citizenship Report*, it announced that it had cut off funding to only nine other groups, leaving a total investment of US$2 million to be divided among thirty-seven groups active in denying, delaying, minimizing, or equivocating about the effects of global warming.

Greenpeace's Davies was encouraged, though, that Exxon actually acknowledged in its 2008 report that it had been financing activities that were, at the very least, unhelpful. Exxon said, "In 2008 we will discontinue contributions to several public policy interest groups whose position on climate change could divert attention from the important discussion on how the world will secure the energy required for economic growth in an environmentally responsible manner." This, however, still contains the euphemistic reference to "groups whose position on climate change could divert attention." It makes the idea of fooling the public about climate science seem almost accidental—and completely innocent.

Even as CEI was losing its Exxon funding, however, a new team leader was emerging in the climate debate: the Heartland Institute. On the surface, Heartland is not quite as credible as the CEI. For example, the aforementioned Heartland "Smokers' Lounge" gives the organization away as one that will make common cause with industrial forces (big tobacco) that have been publicly discredited. The institute used to acknowledge the funders for this sort of work (Philip Morris prominent among them), but found that "critics who couldn't or wouldn't engage in fair debate over our ideas found the donor list a convenient place to find the names of unpopular companies or foundations, which they used in ad hominem attacks against us."[3]

For the record, I'd argue that it would be fair to say that smoking is a dangerous habit that drives up social and health costs—a point accepted by U.S. courts. I would further argue that if someone is taking money from Philip Morris to make a contrary argument, the public has a right to know about the payment before making a finding as to Heartland's credibility.

Early evidence that Heartland's fund-raisers had struck oil came in 2007 with a major newspaper advertising campaign urging a debate between Al Gore and the prominent U.K. climate change denier Christopher Walter—or, as he prefers to be known, the Third Viscount Monckton of Brenchley. Lord Monckton, who styles himself as a former science advisor to then–U.K. prime minister Margaret Thatcher (he appears to have been a junior researcher in the Prime Minister's Office) has no training whatever in science, but has turned his degree in classics and diploma in journalism to full-time denial, pumping out JunkScience.com products like the DVD *Apocalypse? No!* and writing purportedly scientific reports for organizations such as the Science and Public Policy Institute.[4]

Heartland followed up its 2008 ad campaign with the first "International Conference on Climate Change." Once again, Heartland was soliciting compliant scientists, offering an all-expenses-paid trip to New York and a US$1,000 honorarium to any scientist willing to help "generate international media attention to the fact that many scientists believe forecasts of rapid warming and catastrophic events are not supported by sound science, and that expensive campaigns to reduce greenhouse gas emissions are not necessary or cost-effective." And lest there be any confusion as to the intended audience for this discussion, an invitational letter from Heartland senior fellow James M. Taylor noted that elected officials were invited to apply for "scholarships" covering all costs. In March 2009 Heartland launched the second such conference, with a speakers list featuring politicians, science "experts" like Fred Singer and Patrick Michaels, and non-academic "experts" like Christopher Walter.[5]

I would not want to leave the impression from any of the foregoing that I believe think tanks are all bad. There are many admirable institutions that conduct excellent research and advance the discussion of public policy in a way that clearly serves the public interest. But reputable think tanks whose goals are in fact to serve the public good should be willing to post their funding sources prominently and should be doubly prepared to stand accountable when they take firm positions that are outside the mainstream, outside their field of expertise, or, especially, at odds with independent research bodies. I can't think of a situation in which a think tank could credibly attack a national academy of science—as CEI, Heartland, and dozens of others have done on this issue.

The example of Heartland, with its support of "smokers' rights," raises one of the important issues on the climate change file. As Heartland CEO Joseph Bast likes to argue, a smoker—when separated from his children, his friends, and his workmates—is endangering only his own life. There is, I suppose, an argument to be made that it is his right to do so. But the people who are blocking or delaying action on climate change are putting the whole human population at risk. If they are doing so out of some sincere, if wrong-headed, belief that climate change really isn't a problem, then their actions might be forgivable. But if they are doing it because their eagerness to post short-term profits overwhelms their judgment or their interest in the public good, then they should be subject to the full dose of public wrath. Either way, they have put themselves forward as experts—as lifeguards—so their responsibility to be well-informed is that much greater and their negligence of or disregard for the science is even more offensive.

[eight]

DENIAL BY THE POUND

Many wrongs don't make a right, but they sound better

Tens of thousands of scientists now say the media and environmental advocacy groups have it all wrong, that global warming is not a crisis. They point to a cooling trend in global temperatures since 2000, past warming and cooling cycles that were not man-made, and new evidence that carbon dioxide is not a very powerful greenhouse gas.

On March 8 to 10, 2009, the 2009 International Conference on Climate Change will provide a platform for scientists and policy analysts from around the world who question the theory that global warming is a crisis.

For more information about this exciting conference or to register online, please visit www.heartland.org or www.global warmingheartland.org...

FROM A HEARTLAND ADVERTISEMENT IN THE *NATIONAL REVIEW*

Tens of thousands of scientists say global warming is not a crisis? Really? Well, the Heartland Institute wouldn't say it if it weren't true in its way. At the bottom of the same ad, Heartland offers a source for this claim: "More than 34,000 scientists have signed a petition saying global warming probably is natural and not a crisis. See the complete list at www.oism.org/pproject."

Type that link into your computer and you'll wind up at the home of the grandiose-sounding Oregon Institute of Science and Medicine—in reality, a farm shed situated a couple of miles outside of Cave Junction, Oregon (population 17,000). In addition to its founder chemist Arthur Robinson, the Oregon Institute lists six faculty members, two of whom are dead and two others of whom are Robinson's twenty-something sons. The Oregon Institute lists no ongoing research and carries no students. Rather, it publishes books on surviving nuclear war and distributes Arthur Robinson's homeschooler package—a creationist-friendly curriculum based on such material as the 1911 *Encyclopedia Britannica*.

Robinson's views on climate change are clear, and clearly out of step: "I think it's important to speak the truth," he told Jeff Goodell in an interview that Goodell conducted for his book *Big Coal*. Robinson continued, "When you start cutting back on coal and oil, what you're really talking about is depriving millions of people in places like Africa access to cheap energy to improve their lives. One of these days, people will start to see global warming for what it is—a thinly disguised scam by corporations, the United Nations, and big environmental groups to reduce the world's population. Speaking as a scientist, I can tell you that most people who tout global warming are liars, and the sooner we recognize that, the better."

This is all too typical of the language that deniers use in condemning the world's greatest scientists. Robinson, who is not personally a climate scientist, is not suggesting that the climate change experts on the Nobel Prize–winning IPCC are merely mistaken. He says that most of them are "liars." Yet he makes no effort to explain their motives or to say which corporations might benefit from reducing global population. He is also careful enough not to attach a specific name to this libel. Instead, for the last ten years he has dedicated a huge amount of his time to the so-called Oregon Petition. The petition was launched in 1999 under the signature of Dr. Frederick Seitz, the scientist whom a Philip Morris executive dismissed ten years earlier as "quite elderly and not sufficiently rational as to offer advice." Robinson's early role in the petition had been to prepare what Seitz would later refer to as a "twelve-page review of information on the subject of 'global warming.'" Titled "Environmental Effects of Increased Atmospheric Carbon Dioxide," the review was coauthored by Robinson's son Noah and Willie Soon, a Harvard-Smithsonian physicist whose work has been funded by the API and the Exxon-friendly George C. Marshall Institute. Neither peer-reviewed nor published in any scientific journal, the article was laid out and printed in exactly the style used for the prestigious journal *Proceedings of the National Academy of Sciences*. It was then distributed with a letter from Seitz, who is prominently identified as a former National Academy president. Not surprisingly, many recipients took this for an official National Academy of Sciences communication, triggering an uproar that resulted in the National Academy issuing a statement on April 20, 2008, clarifying that it was in no way connected to the petition, that the article had never appeared in a National Academy of Sciences journal—or in any journal—and that the National Academy's position on global warming was opposite to what was being suggested in the petition literature.

This criticism notwithstanding, the petition attracted a huge number of signatures—though the actual number and the veracity of the names have never been established. When early critics pointed out that the petition was littered with names like Perry Mason, Michael J. Fox, and (former Spice Girl) Geri Halliwell, Robinson and company insisted that some of the names were legitimate (Perry Mason refers to a Ph.D. chemist, not to the fictional TV lawyer) and that others had been submitted by mischievous environmentalists. But by failing to include any contact information—or even signatory links to academic institutions—the Oregon Institute made it impossible to verify the petition.[1]

That didn't stop people from trying. In its August 2006 issue the magazine *Scientific American* told readers about its own efforts to confirm or debunk the petition:

> *Scientific American* took a random sample of 30 of the 1,400 signatories claiming to hold a Ph.D. in a climate-related science. Of the 26 we were able to identify in various databases, 11 said they still agreed with the petition—one was an active climate researcher, two others had relevant expertise, and eight signed based on an informal evaluation. Six said they would not sign the petition today, three did not remember any such petition, one had died, and five did not answer repeated messages. Crudely extrapolating, the petition supporters include a core of about 200 climate researchers—a respectable number, though rather a small fraction of the climatological community.
>
> None of the "200 climate researchers" had ever published any refereed research that supported their supposed skepticism, but apparently they still felt strongly enough to sign the petition.

Scientific American conducted this review at a time when the petition listed seventeen thousand signatures. Like other would-be scrutinizers, they found the "scientists" were almost all people with undergraduate degrees and no expertise in climatology, no record of research, and no more ready information than the "review" penned by Robinson, Robinson, and Soon. Yet the petition's organizers were sufficiently pleased with their product that they continued to tout it and to add names. Even though they didn't have the weight of evidence, and even though their signatories were not heavyweights in scientific terms, it appeared that the gross weight of all those purported "scientists" again helped demonstrate a scientific controversy that to the casual observer seemed both legitimate and lively. Robinson and company have continued to solicit signatures ever since, launching a major drive in 2007 that helped push the total from under twenty thousand to over thirty thousand names. Now, when people brandish the reports of the IPCC as evidence that the work of more than 2,500 top experts worldwide supports a conclusion that we must be concerned about the human contribution to climate change, the Heartland Institute and all of their supporters hold up the Oregon Petition and say, so what? "Tens of thousands of scientists now say the media and environmental advocacy groups have it all wrong, that global warming is not a crisis."

This science-by-petition argument has a rich tradition, one that continues to birth new petitions at regular intervals. Dr. S. Fred Singer launched the first obvious effort through his think tank/Web site, the Science and Environmental Policy Project (sepp.org), in 1992. The "Statement by Atmospheric Scientists on Greenhouse Warming" was released four months before the "UN Conference on Environment and Development," the so-called Earth Summit in Rio de Janeiro. With forty-seven signatories, mostly weather forecasters and physicists, it was just the first hint of how the strategy would unfold in the years to come.

In the same year four thousand scientists, including seventy-two Nobel Prize winners, signed the "Heidelberg Appeal," which Singer said also took issue with the science of global warming, but which, in fact, didn't mention climate change or any other environmental issue. It merely called for scientific policy based on "scientific criteria and not on irrational preconceptions."

The "Leipzig Declaration," also promoted by Singer and sponsored by the Science and Environmental Policy Project, arose out of a conference in Leipzig, Germany, in 1995 and included a mix of eighty scientists and twenty-five TV weather forecasters. It was rereleased in 1997 and again in 2005. When Øjvind Hesselager of the Danish Broadcasting Company tried to verify the signatories in 1997, he was working with a list of eighty-two, thirty-three of whom were listed as European. In reporting his findings he gave these results:

- He was unable to locate four of the signatories at all;
- He found twelve who denied having signed or, in some cases, ever having heard of the "Leipzig Declaration";
- He found many who were not qualified in fields even remotely related to climate research. They included medical doctors, nuclear scientists, and entomologists; and
- He discovered that two of the signatories had financial ties to the German coal industry or the government of Kuwait (Robert Balling and Patrick Michaels).[2]

A disturbingly successful Canadian petition was launched in 2006: sixty "accredited experts" published an open letter to Prime Minister Stephen Harper in the April 6 edition of the *Financial Post,* dismissing as meaningless any argument that "climate change is real" and insisting that "allocating funds to 'stopping climate change' would be irrational." This list of sixty included a full selection of the usual suspects drawn from all over the world, but thanks to a well-funded rollout the letter received a surprising amount of positive news coverage. The strategy really showed its value a week or two later when an outraged group of Canada's foremost climate scientists responded with a letter of their own, demanding that Prime Minister Harper set aside the list of sixty and start taking the issue of climate change seriously. Well-intentioned though the second letter clearly was, its principal effect was to reinforce the notion that there was a lively scientific debate. Tom Harris of the Natural Resources Stewardship Project later drove the point home with a national radio advertising campaign touting the letter from the original sixty.

The first letter, however, inspired the DeSmogBlog to begin building a database of the people involved. Our researchers found that only twenty of the scientists on the list were Canadian. The remainder were international players, including people like Fred Singer and Arthur Robinson, whose connection to the denial movement or the energy industry was already well established. A loyal DeSmogBlog reader put together an ExxonSecrets graphic showing the most famous twenty deniers' connections to Exxon-backed think tanks. A few new names made the list, but some, such as the University of Alberta mathematician Gordon Swaters, said later that they had been convinced to sign the petition on the understanding that they were urging the Canadian government to invest more in climate research, *not* that they were denying that climate change is occurring and in need of attention.[3]

Undeterred by the attention, the Heartland Institute sponsored another bigger petition-style attack the next year, this one penned by Dennis T. Avery, a senior fellow at the Hudson Institute. Avery's paper, published on Heartland's Web site, was titled "500 Scientists Whose Research Contradicts Man-Made Global Warming Scares," and it credited the five hundred signatories as "coauthors," implying that each of the five hundred had a hand in Avery's report or, at the very least, signed off on its conclusion. Not.

We got suspicious at the DeSmogBlog, in part because there were so many new names. This was not a collection of people we recognized from previous petitions or serving on the scientific advisory panels of the Exxon think tank brigade. Many of these new scientists also appeared to be respected leaders in their fields. Kevin Grandia set about getting in touch with those for whom we could easily find contact information, asking whether they had participated in Avery's process and whether they challenged the scientific consensus on climate change. Within forty-eight hours we had forty-five responses, all expressing a similar type of outrage. The following are emails received by Kevin Grandia during the last week of April 2008:

> "I am horrified to find my name on such a list. I have spent the last 20 years arguing the opposite."
> —DR. DAVID SUGDEN,
> *Professor of Geography, University of Edinburgh*

> "I don't believe any of my work can be used to support any of the statements listed in the article."
> —DR. ROBERT WHITTAKER,
> *Professor of Biogeography, University of Oxford*

> "I'm outraged that they've included me as an 'author' of this report. I do not share the views expressed in the summary."
> —DR. JOHN CLAGUE, *Shrum Research Professor, Department of Earth Sciences, Simon Fraser University*

And here are two more email messages sent directly to Dennis Avery and copied to Kevin Grandia, again during the last week of April 2008:

> "Please remove my name. What you have done is totally unethical!!"
> —DR. SVANTE BJORCK,
> *Geo Biosphere Science Centre, Lund University*

> "I have NO doubts . . . the recent changes in global climate ARE man-induced. I insist that you immediately remove my name from this list since I did not give you permission to put it there."
> —DR. GREGORY CUTTER, *Professor, Department of Ocean, Earth and Atmospheric Sciences, Old Dominion University*

Despite this reaction, at the time of writing (a year after these complaints were lodged), the original article and the links to the "coauthor" list remain on the Heartland Institute Web site. No apology. No correction. No acknowledgment that the list of five hundred purported deniers is largely if not entirely misrepresented.

Another example of the science-by-petition method fell from the hand of Marc Morano at the United Nations Framework Convention on Climate Change in Poznan, Poland, in December 2008. Morano made his name in Republican circles while still working for Cybercast News Services (owned by the conservative Media Research Center). Cybercast and Morano were the first source in the Swift Boat Veterans for Truth attack launched against John Kerry in the 2004 presidential campaign.[4] Later Morano signed on as the communications director for U.S. Republican Senator James Inhofe from Oklahoma, who, as chair of the Senate Environment and Public Works Committee, received more funding from the oil-and-gas sector than any other senator.[5]

It was Senator Inhofe who invited climate "experts" like the fiction writer Michael Crichton to brief his committee, and it was Morano who set up the meetings and made sure the committee Web site was well-supplied with stories and reports denying the likelihood of climate change. By 2008, however, the Democrats were in control of the committee (California senator Barbara Boxer is the new chair), and Inhofe was left as the ranking minority member on the committee. Morano, however, continued his Web-based activities, and he was able to hitch a ride as part of Boxer's "staff" when the senator attended the international climate conference in Poznan.

In Poland, on December 10, 2008, Morano began distributing a news release announcing, "More Than 650 International Scientists Dissent over Man-Made Global Warming Claims."

The short list of "experts" attached to the release didn't include any of the names that have grown recognizable as participants in the think tank campaign. But a look at the complete list showed many familiar names (Fred Singer, Tim Ball, Sallie Baliunas). Fred Seitz was there too, though he had died more than a year previous—as were three or four other "senior scientists" who were no longer alive to agree with Morano or to correct the record.

The beauty of this tactic as a method of keeping the debate alive is that none of these "scientists" ever have to conduct any actual research or put their views forward to be tested in the scientific peer-review process. They don't even have to be experts in a related field. And they certainly don't have to win the argument. As long as groups of scientists are seen to be disagreeing, the public continues to assume that the science is uncertain. That's why this entire conversation occurs in mainstream media rather than in scientific journals, where you have to prove the veracity of your argument to the satisfaction of a panel of experts in your field. On newspaper opinion pages you can say what you like, knowing that the editor has no relevant expertise. And if you are called on a mistake after the fact, you can insist, as the *Washington Post* did in defense of George Will during a controversy that arose in February 2009, that you are just standing up for free speech.

Will had written, incorrectly, that global sea ice was advancing, when it is in fact in retreat. When critics, both internal and external, called for a correction, the *Post*'s editorial page editor Fred Hiatt said this to the *Columbia Journalism Review:* "It may well be that he is drawing inferences from data that most scientists reject—so, you know, fine, I welcome anyone to make that point. But don't make it by suggesting that George Will shouldn't be allowed to make the contrary point. Debate him." Critics responded that this is not a matter of opinion; it's a point of fact. But Will didn't apologize or correct the mistake, and the *Post* stood by him.

The target audience for the endless flow of "petition science" isn't the community of climate scientists. The target audience is the increasingly confused general public, and politicians who are either similarly bewildered or seriously beholden to special interests. As long as people believe that the science of climate change is uncertain, still a topic of legitimate debate in the scientific community, they will shrink from supporting policies that demand or even suggest a significant change in habit. Few people will want to give up their car or spend money retrofitting their home heating system if they believe that scientists are still arguing over the truth of global warming. And few politicians are going to risk introducing dramatic policy proposals to change the way society treats fossil fuels when the public is skeptical that this is necessary. So Arthur Robinson and company never have to win this argument. If they wanted to, they would have to act like scientists, spending their time in the lab and writing papers in science journals, rather than acting like public relations people—spending their time on the speakers' circuit and writing opinion-page articles in venues where every fact is open for debate.

[*nine*]

JUNK SCIENTISTS

An expert for every occasion; an argument for every position

The notion of a "junk scientist" is necessarily pejorative. It suggests that some science is "sound"—that it is based in good research and excellent practice—and other science is slovenly, suspect, and unreliable. My scientist friends don't make a distinction. To them, there is only science—pure and evidence-based, even when it evolves as our scientific knowledge and understanding increase. But there are people who have made a career of the discussion of "junk science." Primary among these is Steven Milloy, introduced in Chapter 4 as an early executive director of TASSC, a coauthor of the American Petroleum Institute's "Global Climate Science Communication Action Plan," and the proprietor of the Web site JunkScience.org.

But Milloy isn't a scientist and doesn't present himself as such. He is a public relations person, a lobbyist, and a sometimes-journalist. He is employed as a regular columnist on the Fox News Web site and is periodically featured in Fox TV coverage as a "junk science expert"—a title they give him without ever mentioning that he is a registered lobbyist for many companies whose businesses would be affected by regulations governing, say, greenhouse gases or genetically modified foods.

By my definition, junk scientists are people like Dr. Fred Seitz, who leverage their scientific credentials in order to speak out as scientific experts even when the topic on which they are speaking is outside their field of expertise. These are the kind of people whom the API action plan targeted for identification and media training, people willing to throw themselves into the middle of the public debate about climate change whether or not they are trained in the field or currently engaged in climate change research.

Take Dr. Benny Peiser as an example. Peiser is a senior lecturer in the School of Sport and Exercise Sciences at Liverpool's John Moores University. He is a social anthropologist whose undergraduate work was in English studies and sports science. If you wonder why, of the thirty-thousand-odd signatories to Arthur Robinson's petition, I would single out Benny Peiser, it is because he opened up hostilities in January 2005 by attacking Dr. Naomi Oreskes and attempting to discredit her study on climate change consensus, "Beyond the Ivory Tower: The Scientific Consensus on Climate Change," published in *Science* the previous month.

As mentioned in Chapter 1, Oreskes had searched the databases of published peer-reviewed science for articles on "global climate change," and found that of 928 articles published between 1993 and 2003, none disagreed that human activity was causing climate change in the world. Less than a month after *Science* published Oreskes's article, Peiser weighed in with a serious complaint. He wrote *Science* to say that a search of the literature reveals not 928 articles but more than 12,000. He asked why Oreskes had limited her sample: "What happened to the countless research papers that show that global temperatures were similar or even higher during the Holocene Climate Optimum and the Medieval Warm Period when atmospheric CO_2 levels were much lower than today; that solar variability is a key driver of recent climate change, and that climate modeling is highly uncertain?" *Science* did not publish Peiser's letter, so he posted it to his Web page at John Moores University.

Score a point for Peiser: while Oreskes said in her original article that she had used the search term "climate change," she had actually specified "*global* climate change," and thanks to Peiser's intervention, she had to clarify this "error" in a later edition of *Science.* Oreskes also searched only for peer-reviewed documents—that was her whole point—while Peiser had thrown the net wide enough to include all articles, peer-reviewed or otherwise. He therefore wound up with a larger sample, and he reported finding "34 abstracts reject or doubt the view that human activities are the main drivers of 'the observed warming over the last 50 years.'"

Peiser had actually submitted a contradictory paper, which *Science* rejected as too long for a correction, inviting him to submit a letter to the editor. This he did, setting in motion a further fact-checking flurry that lasted almost two more years. Scientists around the world demanded to see these thirty-four abstracts. If legitimate, peer-reviewed studies contested the consensus view on climate change, no one else could find them, so they asked Peiser to share his sample.

A leading interrogator in this period was Tim Lambert, a computer scientist at the University of New South Wales in Australia. Lambert writes an amazingly articulate and often-courageous blog called Deltoid (scienceblogs.com/deltoid), specializing in exposing scientific silliness, especially the errors, irrelevancies, and inaccuracies of those who challenge the science of climate change. After several months of Lambert asking,

Peiser finally provided the list, which Lambert posted—commenting on the obvious: contrary to Peiser's claims, none of the abstracts took specific issue with the global climate consensus.

Peiser refused to back down, insisting that some of these papers must surely be "ambiguous" at the very least and claiming that he still had a paper in hand that contradicted the global climate science community. And he did. After sixteen months of further nagging, he provided the particulars and finally admitted that this single contrarian piece was, in fact, not peer-reviewed. It was an opinion piece printed in the journal of the American Association of Petroleum Geologists.

You might well ask, who has the time to keep track of all this stuff, anyway? And then you might take a moment to think a kind thought about Tim Lambert for his persistence. You might also ask, what difference does it all make? And then you might shudder. Because while *Science* reacted gingerly to Peiser's criticism, the think tank echo chamber was bouncing his erroneous information around the Internet and spilling it into mainstream media reports as frequently as possible. The most famous citation of Peiser's work occurred on September 5, 2006, in a speech ("Hot & Cold Media Spin: A Challenge to Journalists Who Cover Global Warming") that Senator James Inhofe delivered on the floor of the U.S. Senate. Although Peiser had admitted almost a year earlier that his thirty-four abstracts were shaky at best and fictional at worst, the Senate's foremost recipient of campaign funding from the oil-and-gas industry presented Peiser's work as a standing devastation of Oreskes's earlier work:

> On July 24, 2006, the *Los Angeles Times* featured an op-ed by Naomi Oreskes, a social scientist at the University of California San Diego and the author of a 2004 *Science* magazine study. Oreskes insisted that a review of 928 scientific papers showed there was 100 percent consensus that global

> warming was not caused by natural climate variations. This study was also featured in former vice president Gore's *An Inconvenient Truth.*
>
> However, the analysis in *Science* magazine excluded nearly eleven thousand studies, or more than 90 percent of the papers dealing with global warming, according to a critique by British social scientist Benny Peiser.
>
> Peiser also pointed out that less than 2 percent of the climate studies in the survey actually endorsed the so-called "consensus view" that human activity is driving global warming, and some of the studies actually opposed that view.

There is no record of Senator Inhofe ever correcting the record or apologizing to the Senate for promoting Peiser's work long after it was proved to be inaccurate.

No warning is attached to a Ph.D., and no gradations exist to suggest that one Ph.D. is relevant and another less so. For example, there is no quick and easy marker showing that Benny Peiser claims credit to a single book chapter on climate change, in the 2003 book *Adapt or Die*, or that his lifetime list of peer-reviewed publications on all topics amounts to three—none of them in climate science. In his spare time he is an advising member of the U.K. Scientific Alliance, an organization that was started by gravel-pit owner Robert Durward in a fit of anger about "all this environmental stuff."[1]

But if Peiser's bona fides are not immediately in evidence, you don't have to search far for his name. He's on the Canadian list of 60 scientists quoted in the previous chapter. He's on Marc Morano's list of 650. And he was on the Heartland Institute's guest list as a speaker at the "2nd International Conference on Climate Change."

For the record, Naomi Oreskes is incredibly accomplished and widely admired. She received her B.Sc. in mining geology from the Royal School of Mines, Imperial College, University of London, in 1981. She taught and conducted research in geology, philosophy, and applied earth sciences at Stanford beginning in 1984 and received her Ph.D. in the Graduate Special Program in Geological Research and History of Science from that prestigious school in 1990. She also won a National Science Foundation Young Investigator Award in 1994. Today she is provost of the Sixth College at the University of California, San Diego. And for three years at least, every time she was quoted or interviewed anywhere in the world, Peiser would pop up offering a counterpoint, even though he could not prove his claim.

IF ORESKES HAD not discounted papers that were later proved to be incorrect, she might have listed one paper challenging the consensus on climate change. It was published in 2003 by the astrophysicists Sallie Baliunas and Willie Soon in the journal *Climate Research.* Titled "Proxy Climatic and Environmental Changes of the Past 100 Years," the paper gathered up the work of more than a dozen other climate scientists and concluded, "Across the world, many records reveal that the 20th century is probably not the warmest nor a uniquely extreme climatic period of the last millennium."

The scientists cited in the paper reacted first in confusion and then in anger. They argued that Baliunas and Soon had misrepresented or misinterpreted their data, they wrote hasty counterpoints, and they issued a news release through the American Geophysical Union reinforcing the validity of their earlier work. Ultimately, the publisher of *Climate Research,* Otto Kinne of Inter-Research, wrote a note in a subsequent edition of the journal, saying that "CR [*Climate Research*] should have been more careful and insisted on solid evidence and cautious formulations before publication" and that "CR should have requested appropriate revisions of the manuscript prior to publication."

But news got out afterward that Kinne wrote the critical note only after half the editorial board of the journal had resigned in protest over how the Baliunas article had been handled.

Baliunas, though a much more impressive scientist than Peiser, is still, interestingly, out of her element in climate science. She has published more than two hundred peer-reviewed papers in prestigious astrophysical journals, mostly on the sun and sunlike stars. Lately she has added half a dozen publications vaguely related to climate change, primarily in second-tier journals such as *Energy and Environment.* Yet despite—or maybe because of—the dustup over her CR paper, Senator James Inhofe chose her to brief the Senate Environment and Public Works Committee.

Baliunas has also enjoyed an increase in income and attention since emerging as a climate contrarian. Aside from accepting research funding directly from the American Petroleum Institute, Baliunas is also listed on ExxonSecrets as a former board member and advisory board chair at the George C. Marshall Institute; environment/science editor at Tech Central Station; a past contributing editor to the Western Fuels Association publication *World Climate Report;* a member (with Benny Peiser) of the advisory board of Robert Durward's U.K. Scientific Alliance; a former expert for the Competitive Enterprise Institute; and a science expert working on behalf of the National Center for Public Policy Research and the Hoover Institution. All of the foregoing (except the Western Fuels Association *World Climate Report* and the gravel magnate's Scientific Alliance) are on record as having accepted funding from ExxonMobil.

ONE OF THE most recognizable names among the actual climate scientists who continue to downplay the risk of global warming is that of Patrick Michaels, a frequent spokesperson on behalf of the coal industry since the days of the Western Fuels Association ICE campaign. We know about Michaels's history in part because of Ross Gelbspan's landmark books exposing the climate confusion campaign, *The Heat Is On* (1998) and *Boiling Point* (2005). Gelbspan had documented as early as the mid-1990s that scientists including Michaels, Sherwood Idso, Robert Balling, and Fred Singer were carting away tens of thousands—sometimes hundreds of thousands—of dollars from sources ranging from the Western Fuels Association to the British Coal Corporation, the German Coal Mining Association, OPEC, and the Kuwait Foundation for the Advancement of Sciences.

By 2005, when we were working to launch the DeSmog Blog, Ross Gelbspan had done so much reporting on the fossil fuel funding to contrarian science that he thought the story was over. He told me at the UN Framework Convention on Climate Change meeting in Montreal in December 2005 that the public and the politicians already knew that people like Pat Michaels and Sallie Baliunas were no longer credible. Surely, Gelbspan said, it was time to move on—to start talking about solutions.

Not quite. Six months later a sputtering Gelbspan was on the phone, reporting the kind of scoop that he had hoped would no longer be relevant. Someone had leaked him a memo, now posted on DeSmogBlog, from the Intermountain Rural Electric Association—a cooperative comprising mostly coal-fired electrical generating utilities. The memo reads like a script from the old ICE campaign: "Al Gore and others state that the scientific community has reached a consensus and that the debate is over. That is simply not true. Disputing this contention are climatologists, meteorologists and astrophysicists like Richard Lindzen, William Gray, Fred Singer, Roy Spencer, Patrick Michaels, Robert Balling, and Craig Idso and the 17,000 signatories to Arthur B. Robinson's Petition Project."

Here again we see someone questioning the risk of global warming, not by criticizing the science, but by lining up a list of contrarian scientists and insisting that this constitutes a legitimate debate. We also see evidence of what appeared to be a continuing effort to rework the meaning of the word "consensus," which the *Oxford English Reference Dictionary* defines as "general agreement"—not unanimity. The memo's author, Intermountain Rural Electric Association general manager Stanley Lewandowski, then went on to urge his members to join the public debate: "I am enclosing copies of a fact sheet that, if you are inclined, you could print copies for your employees and ask them to mail these to friends, relatives and acquaintances. The information could also be used for informing your members, the local media and local and state elected officials. We plan to contact unions, other social and business groups, as well as industrial corporations served by the investor-owned utilities. We cannot allow the discussion to be monopolized by the alarmists."

Lewandowski writes about the pro–carbon dioxide ad campaign produced by the Competitive Enterprise Institute and financed by GM and Ford. He talks about the involvement of coal-consuming giants including American Energy Power and the Southern Company, and he notes the great cooperation that he and his collaborators are getting from the National Association of Manufacturers. Finally, he notes that his association has just paid Pat Michaels US$100,000, and he urges others to do the same: "We . . . believe that it is necessary to support the scientific community that is willing to stand up against the alarmists and bring a balance to the discussion."

Recognizing all the talking points from the mid-1990s Western Fuels Association campaign, Gelbspan couldn't believe that the industry was still doing the same thing a decade later, shamelessly and transparently (although Stanley Lewandowski likely had not written the memo thinking that it was going to be quite so widely circulated). Dubbing his scoop the "Vampire Memo," Gelbspan said he was obviously wrong about the campaign of denial being at an end. "Apparently," he said, "it just won't die."

THE SUCCESS OF the junk science campaign rests on three factors. First, the "science" conversation most often occurs outside the institutions of science. The Intermountain Rural Electric Association isn't funding Pat Michaels to go back into his lab and do research helping the world to a better understanding of how human activities are affecting the climate. The coal-fired-utility owners are paying him to "stand up against the alarmists and bring a balance to the discussion." And again, Sallie Baliunas, though congratulated in certain circles for having published a much-maligned contrarian paper in a peer-reviewed journal, is employed by Exxon service providers such as Tech Central Station and the National Center for Public Policy Research not to pursue research in astrophysics, but to speak about global warming.

The second factor contributing to the success of this public relations campaign is the skill and talent of the practitioners. If you google videos with people like Robert Carter from James Cook University in Australia or Tim Ball, once a professor at the University of Winnipeg in Canada, you will see charming and entertaining men, practiced lecturers who can string together a good story and stand up to a bit of media grilling. Knowing how difficult that can be—knowing from a public relations perspective how much time goes into preparing senior executives to talk to crowds or face the media—it seems likely that, just as the Western Fuels Association ICE campaign suggested, these people have been selected for their skills or trained for the task.

The third factor that contributes to the junk science campaign's effectiveness is something known in the public relations business as an echo chamber. That's the reverberating network of think tanks, blogs, and ideologically sympathetic mainstream media outlets that distribute and circulate contrarian information.

For no issue has the echo chamber been used more effectively than in the "debunking" of the so-called hockey stick graph by the amateur climate science expert Stephen McIntyre. I use the term "amateur" not as a pejorative but to indicate that McIntyre, a retired mining executive and an investor, is not a professional scientist. Nor is he so obviously employed by industry-funded think tanks, though ExxonSecrets lists him as a George C. Marshall Institute "expert." But McIntyre has brought a dogged professionalism to his criticism of certain very narrow points of climate science and in the process has made himself famous in denier circles.

Working with a University of Guelph, Ontario, economist named Ross McKitrick, McIntyre launched an attack in 2003 on what has come to be known as the Mann hockey stick graph—a reconstruction of Earth's temperatures over the last one thousand years. Created by a team led by Pennsylvania State University paleoclimatologist Michael Mann, it was one of the principal images used in the *Summary for Policymakers* from the Third Assessment Report of the IPCC. Perhaps obviously, it comes out in the shape of a hockey stick, with temperatures running relatively flat for nine hundred years or more (the stick) and then spiking up in the latter half of the 20th century (the blade).

In a full frontal assault published in the journal *Energy and Environment,* McIntyre and McKitrick said that the graph "contains collation errors, unjustifiable truncation or extrapolation of source data, obsolete data, geographical location errors, incorrect calculation of principal components and other quality control defects."

Energy and Environment, however, is a less than prestigious journal; its editor, Sonja Boehmer-Christiansen, is another member of the Robert Durward U.K. Scientific Alliance. So McIntyre and McKitrick continued to press for more high-profile attention. They were rebuffed by *Nature* but succeeded in 2005 in getting a second version of the paper published in the journal *Geophysical Research Letters.*

The criticisms, though actively rebutted by Mann and others, still forced Mann to add a clarification to the record—a "victory" that gave McIntyre and company courage enough to claim that the hockey stick had been discredited. And they didn't stop there. As the news passed through the think tank echo chamber, the assault on the hockey stick became a proxy for an attack on all climate science. There was an error in the hockey stick, so (the think tanks argued) that cast all climate science into doubt. By the time the message was spun back out to the most malleable of mainstream media outlets, the point seemed conclusive. Barry Cooper (that University of Calgary professor who set up the oil-industry slush fund to finance the Friends of Science), dismissed all "global warming hysteria" in a *Calgary Herald* column on the last day of 2008 ("Cooling Global Warming Hysteria Just One Story of '08"), sneering especially at "the notorious intellectual swindle of the hockey stick graph." Lawrence Solomon, writing in Canada's *National Post* on February 6, 2009 ("Mann's Conclusion Not to Be Believed"), said, "Conclusion about the hockey stick graph: Mann-made science does not support the hypothesis that global warming is man-made."

The hockey fight also carried over to the political arena. Senator Inhofe, in his all-encompassing "Hot & Cold Media Spin" climate change speech on the Senate floor in 2006, gave the distinct impression that the "fall" of the hockey stick was a matter of established fact: "The 'hockey stick' was completely and thoroughly broken once and for all in 2006. Several years ago, two Canadian researchers tore apart the statistical foundation for the hockey stick. In 2006, both the National Academy of Sciences and an independent researcher further refuted the foundation of the 'hockey stick.'"

Well, that was Inhofe's interpretation of the National Academy of Sciences review. For another interpretation you could refer to *Nature,* one of the two most prestigious scientific journals in the world, which carried a June 2006 account of a National Academy of Sciences panel review of Mann's work under the headline, "Academy Affirms Hockey-Stick Graph."

Statistician Edward Wegman completed a second, Congress-ordered review of the hockey stick in July 2006. Wegman complained that Mann was not trained in statistics and had not sought the assistance of a statistician in preparing the graph. He endorsed some of McIntyre's criticisms and concluded, "Overall, our committee believes that Dr. Mann's assessments that the decade of the 1990s was the hottest decade of the millennium and that 1998 was the hottest year of the millennium cannot be supported by his analysis."

So the debate rages on. But here's the suspicious part: no one among the ideological media commentators or among industry's favorite political leaders seems to be asking any searching questions about the science behind the stick. They seem uncurious about whether Mann's work has been tested by other scientists or confirmed or falsified by the use of other methods or other proxy data sources. They just want to whip up and sustain the controversy around this single piece of evidence.

Here's why: if you look at the other climate-reconstruction graphs that have been prepared and published since Mann's leading example, you could outfit a whole hockey team and still have sticks left over. As we detail at DeSmogBlog, more than a dozen other scientists have come up with numerous reconstructions, which all feature a long, relatively straight line through most of the past millennium punctuated by a sharp upturn in the middle of the last century. McIntyre and Wegman can quibble with Mann's methodology or wiggle the historic figures however much they like, but they can't challenge the slope of the "blade." And they seem unable to find a single foible or flaw in the content or methodology of all these other studies.

It's clear from reading McIntyre's blog, ClimateAudit.org, that he puts huge amounts of time and a significant amount of care into what he does. He also gets the odd thing right—he caught another statistical error in the past year in the way that the U.S. National Oceanic and Atmospheric Administration was reporting temperatures. But you can also tell that the argument between McIntyre and Mann has grown deeply personal, and that McIntyre is not the least interested in resolving the underlying issue. He's famous because he is a think tank darling, and he seems more interested in maintaining that fame than in really answering whether the world is burning while he fiddles.

It should be remembered, too, that Michael Mann is one of the most impressive climate scientists in the country. He is an associate professor at Pennsylvania State University and the director of Penn State's interdepartmental Earth Systems Science Centre. He has more than eighty peer-reviewed articles to his credit, just one of which availed itself to a very narrow technical criticism from Steve McIntyre. Mann is also one of the founders of the Web site RealClimate.org, one of the best sources on the Internet for clear and accessible, but soundly scientific, discussion about climate change.

IF, BY DINT of his doggedness, Steve McIntyre stands as one of the more credible climate change quibblers, Christopher Walter, the Third Viscount Monckton of Brenchley, must register a little further down the list. Styling himself "Lord Monckton" (or, if you're a close friend or a Canadian talk-radio host, "Lord Christopher"), the Viscount is forever inserting himself into the climate conversation, on the side of confusion and disbelief.

His actual science credentials are thin. His most prominent current position seems to be as "chief policy advisor at a think tank called the Science and Public Policy Institute." (This Frontiers of Freedom knockoff, founded with a start-up grant from Exxon, has my favorite denier webname: SPPINstitute.org. You have to give them points for transparency.) Monckton's biography on the SPPINstitute reads as follows: "Christopher, Third Viscount Monckton of Brenchley, was Special Advisor to Margaret Thatcher as U.K. Prime Minister from 1982 to 1986, and gave policy advice on technical issues such as warship hydrodynamics."

This is unfounded self-promotion. In 1982 Monckton was a thirty-year-old researcher who had studied classics and journalism. You have to wonder if he would have been the expert of choice when the Iron Lady and the Royal Navy were considering the design of their warships. The biography goes on to say that Monckton had special skills in "psephological modelling (predicting the result of the 1983 General Election to within one seat)." Those who have served in political offices during elections might speculate that this meant Monckton had won the office election pool.

But if Monckton is to be suspected of overstating his qualifications, his claim to Nobel status must certainly be his most outrageous exaggeration. The SPPINstitute biography goes on to say, "His contribution to the IPCC's Fourth Assessment Report in 2007—the correction of a table inserted by IPCC bureaucrats that had overstated tenfold the observed contribution of the Greenland and West Antarctic ice sheets to sea-level rise—earned him the status of Nobel Peace Laureate. His Nobel prize pin, made of gold recovered from a physics experiment, was presented to him by the Emeritus Professor of Physics at the University of Rochester, New York, USA." This is an insult to the members of the IPCC, whom the Nobel committee actually honored along with Al Gore. Official Nobel recipients—the actual authors and editors of the IPCC report—received a very handsome certificate (my friend, University of Victoria climate modeller and leading IPCC chapter author Andrew Weaver, described his with considerable pride). With Monckton not being on the Nobel list, it is beyond comical that one of his buddies would melt down the components of an old experiment to fashion a pin. It's worse yet that Monckton would boast of this as somehow legitimizing his claim to Nobel status.

Monckton says in a 2006 letter to two U.S. Senators that he is a Member of the House of Lords. But the evidence, described in Tim Lambert's Deltoid blog, shows that he is not. Monckton (or someone using his computer) tried to convince the editors at Wikipedia that he had won a £50,000 libel judgment against *Guardian* columnist George Monbiot. But the evidence, available at Monbiot.com, shows that he did not. Most broadly, Monckton says that former U.S. vice president and legitimate Nobel laureate Al Gore is all wrong about climate change. But the evidence . . . the evidence is such a nuisance.

The clash of fact and comedy came to a head in July 2008 after Monckton submitted to *Forum on Physics and Society* (an online newsletter of the American Physical Society) a lengthy attack on the most recent IPCC report. Almost immediately the news began pounding through the think tank echo chamber that one of their own had published a "peer-reviewed" article contradicting the scientific consensus about climate change.

Unless you use the definition of "peer" that invokes membership in the British peerage, this claim fell down rather publicly. The newsletter had edited Monckton's piece. They had even assigned a real physicist to help him make sense of the math. But the *Forum on Physics and Society* is not a refereed journal, and Monckton's piece had neither enjoyed nor endured a formal process of review. In response to complaints from American

Physical Society members who were outraged that Monckton was using their organization to claim credibility for his views, the society quickly posted this note on top of Monckton's harangue, making it clear that the organization had not moved from its support of the consensus and that Monckton's piece was not considered a formal scientific paper: "The following article has not undergone any scientific peer review, since that is not normal procedure for American Physical Society newsletters. The American Physical Society reaffirms the following position on climate change, adopted by its governing body, the APS Council, on November 18, 2007: 'Emissions of greenhouse gases from human activities are changing the atmosphere in ways that affect the Earth's climate.'"

Monckton was incensed. He wrote what the SPPINstitute called a "sharp letter" to American Physical Society president Arthur Bienenstock, condemning the advisory as "discourteous" and demanding the names and addresses of everyone who was involved in adding it above his article. This excerpt from Monckton's July 18, 2008, letter of complaint, now available at scienceandpublicpolicy.org, is typical of the Viscount's usual writing style: "If the Council (of the American Physical Society) has not scientifically evaluated or formally considered my paper, may I ask with what credible scientific justification, and on whose authority, the offending text asserts *primo,* that the paper had not been scientifically reviewed when it had; *secundo,* that its conclusions disagree with what is said (on no evidence) to be the 'overwhelming opinion of the world scientific community'; and, *tertio,* that 'The Council of the American Physical Society disagrees with this article's conclusions'?"

It seems certain that Monckton is not a fan of George Orwell, who would have steered the Viscount away from the Latin numbering. In "Politics and the English Language," Orwell says: "Bad writers, and especially scientific, political, and sociological writers, are nearly always haunted by the notion that Latin or Greek words are grander than Saxon ones."

Despite his literary ticks, despite his Orwellian track record of rewriting history and his paucity of scientific credentials, he remains a favorite on the lecture and radio talk show circuit. His video from a lecture titled "Apocalypse? No!" is available on the SPPINstitute Web site or from Steven Milloy at JunkScience.com. He was also on the speakers list at the Heartland Institute's "International Conference on Climate Change."

A LAST CATEGORY of junk scientist is the unwitting accessory. The unwitting accessory is not to be confused with the witless denier—the kind of person who is armed with too little information and too great a sense of personal certainty. Rather, these are people who are dragged into the argument without their permission and sometimes without their knowledge. For example, Dennis Avery took advantage of dozens of unwitting collaborators in compiling his list of "500 Scientists Whose Research Contradicts Man-Made Global Warming Scares." But an even creepier example cropped up in a video that the Heartland Institute sent to eleven thousand Canadian schools, urging teachers to tell their students that scientists are exaggerating how human activity is affecting global climate.[2]

The DVD, now on YouTube, was titled *Unstoppable Solar Cycles: The Real Story of Greenland*, and it begins with Rie Oldenburg, curator of the Narsaq Museum in Greenland, talking about the warm and habitable climate that made the early Viking settlement in Greenland thrive. The video then switches suddenly from an interesting historical retrospective to a political polemic. One minute you have people talking about early Viking fashions, and the next, the veteran denier Willie Soon is talking about coping with "natural changes in the Earth's climate system." What began as a charming history story is suddenly a video tribute to Fred Singer and Dennis Avery's latest book, *Unstoppable Global Warming: Every 1500 Years*. It becomes a carefully constructed argument to convince children, in the words of the announcer, that "this current (and natural) climate cycle is what most people call 'global warming.'"

When someone passed a copy of the DVD to Richard Littlemore, he was struck by the two-videos-in-one effect and curious what the Greenland participants had been told. So he phoned Rie Oldenburg. Her response: "I am somewhat horrified." The effects of climate change are more evident in Greenland than almost anywhere else on Earth, and Oldenburg had just helped convene an environmental seminar intended to call people's attention to the dangers and explain the human causes. As to the video, far from being forthright about their purpose, she said the producers had called on short notice and told her they were working on a film about Viking history. Although they provided nothing in writing as to their intentions, they said they were shooting for the Discovery Channel.

The real client was the Idea Channel, an advocacy organization that exists to promote the free-market economic theories of Milton Friedman. Idea Channel spokesperson Christina Belski said the video was funded by a private donor and by the Heartland Institute, which was distributing it to high schools in Canada and sending it out, unsolicited, to other individuals. It seems revealing, though, that while they had budgeted to send eleven thousand unsolicited copies to schools and other individuals, they couldn't scrape up postage to send a copy to Oldenburg, or even to send her an email advising that it was available for viewing on YouTube. It's almost as though they had something to hide.

[*ten*]

FROM DENIAL TO DELAY

A more reasonable—and more dangerous—trend in obstructing action on climate change

As the evidence of climate change has become more compelling—as the science has grown more certain and as people have come to recognize the changes occurring before their very eyes—a new and more dangerous form of junk scientist has begun to emerge: the nondenier deniers. These are people who put themselves forth as reasonable interpreters of the science, even as allies in the fight to bring climate change to the public's attention. But then they throw in a variety of arguments that actually undermine the public appetite for action. They say, for example, that it's too late to stop the warming, so we should concentrate on adaptation. They suggest that warming may not be bad, and might even be good. They argue that we must "balance" action on climate change with concern for the economy. And one bait-and-switch specialist suggests that we should withhold spending on climate change mitigation as long as there are other, arguably higher priorities, such as poverty,

HIV, or malaria, that remain unfunded. This is Bjørn Lomborg, the Danish game theorist whom we met in Chapter 1.

Lomborg's debating style emerged in his 2001 best-seller, *The Skeptical Environmentalist*—a hymn to industrialization, which argued that the environment was not getting worse, it was getting better. Papering over evidence to the contrary, Lomborg compiled a whole volume of anecdotes and statistics that appeared to show how our technological mastery of nature had improved the world in which we live. He concentrated especially on the degree to which air and water quality improved in the industrialized world between the early 1970s and the late 1990s.

This example illustrates one of Lomborg's first lapses in logic, one common to many antigovernment, antiregulation campaigners. Lomborg and his industry supporters frequently use the success of old environmental regulations to argue that new environmental regulations are unnecessary. They say we don't need to worry about air pollution now because the private sector has already shown that it can solve the problem: you can, for example, see across California's San Fernando Valley with a frequency that was unimaginable thirty years ago. But it's important to remember that industry didn't make those positive changes of their own volition. Air quality improved in California, and throughout the U.S., because the state of California took the lead in passing emissions-control and antipollution legislation—and many other states followed suit. But when Californians recently tried to take similar initiative by imposing innovative climate change policies, the Bush administration threw up legal hurdles to block progress, and the automakers took the state to court.

For all its reasonable tone, it would be an understatement to say that *The Skeptical Environmentalist* was controversial. So many scientists complained about its content that the Danish Committees on Scientific Dishonesty launched a review of the book and, in a report delivered early in 2003, concluded: "Objectively speaking, the publication of the work under consideration is deemed to fall within the concept of scientific dishonesty . . . In view of the subjective requirements made in terms of intent or gross negligence, however, Bjørn Lomborg's publication cannot fall within the bounds of this characterization. Conversely, the publication is deemed clearly contrary to the standards of good scientific practice."

The Danish Committees cited *The Skeptical Environmentalist* for "Fabrication of data; Selective discarding of unwanted results (selective citation); Deliberately misleading use of statistical methods; Plagiarism; and Deliberate misinterpretation of others' results." Ultimately, however, the committees used Lomborg's own lack of credentials to absolve him of responsibility for his errors. They said they could not find him culpable because he was out of his depth in the fields of science. But if you are interested in an exhaustive account of Lomborg's errors in this and subsequent books, a Danish academic named Kåre Fog keeps a Web site record at www.lomborg-errors.dk.

In an introduction on the Web site Fog poses the question, "Why is it essential to point out the errors?" In answer, Fog says:

> First, because in the handling of errors, Lomborg is not a normal person. A normal person would apologize or be ashamed if concrete, factual errors or misunderstandings were pointed out—and would correct the errors at the first opportunity given. Lomborg does not do that. For example, when *The Skeptical Environmentalist* was heavily criticized in a review in *Nature*, Lomborg's reaction was: "If I really am so wrong, why don't you just document that?"—and then, when this was documented, he ignored the facts.
>
> Second, because you cannot evaluate Lomborg's books just by reading them and thinking of what you read. For

> every piece of information in the books, you have to check if it is true and if the presentation is balanced. If the concrete information given by Lomborg is correct and balanced, then it follows that his main conclusions are also correct. But if the information is flawed, then the main conclusions are biased or wrong. Therefore, in principle, you can only evaluate the books after having checked all footnotes, read all references, and checked alternative sources. This will be a huge task for any reader, but when the errors are described and presented in one place—this Web site—then the task becomes manageable.

Despite the criticisms (Fog documents 319 specific errors, exaggerations, or logical flaws), Lomborg became the toast of the antienvironmental movement, receiving the Julian Simon Award from the D.C.–based Competitive Enterprise Institute and striking out on a perpetual speaking tour in the U.K., the U.S., and, thanks to the Fraser Institute, in Canada.

In his first book, Lomborg minimized or ignored climate change as an issue, but he turned his full attention to the topic in 2004, organizing the first meeting of the Copenhagen Consensus Center. The center, founded by Lomborg and sponsored by the right-of-centre Danish government, posts this as its mission: "The Copenhagen Consensus Center is a think-tank in Denmark that publicizes the best ways for governments and philanthropists to spend aid and development money. We commission and conduct new research and analysis into competing spending priorities—especially those investments designed to reduce the consequences of the world's biggest challenges." It sounds wonderfully noble and urgently necessary in a world where government budgets are tight and priorities many.

In 2004 Lomborg brought together a panel of leading economists, including four Nobel laureates, and asked them to evaluate a selection of the world's problems, weighing the costs and challenges involved in finding solutions and producing a prioritized list of those most deserving of money. In a July 8, 2006, feature story titled "Get Your Priorities Right," the *Wall Street Journal* offered this brief summary as the "well-publicized result": "While the economists were from varying political stripes, they largely agreed. The numbers were just so compelling: $1 spent preventing HIV/AIDS would result in about $40 of social benefits, so the economists put it at the top of the list (followed by malnutrition, free trade and malaria). In contrast, $1 spent to abate global warming would result in only about two cents to 25 cents worth of good; so that project dropped to the bottom."

Lomborg is quite brilliant at presenting this assessment as an inevitable, even unfortunate, bit of news for people who care about the risks of climate change. He argues, with apparent passion, that he also cares about climate change—it's just that he is trapped by the numbers and crippled by the cost of fighting climate change. By his careful economic analysis, he says it just isn't worth diverting valuable human resources from more pressing problems like AIDS, malnutrition, and the provision of fresh water to people in the developing world. You can see the skill with which he makes this argument in a YouTube video of a TED conference presentation. He comes across as charming and sincere, truly disappointed that he doesn't have a budget for climate change remediation, but flatly and irrevocably convinced.

The question, however, is this: if Lomborg were really certain that the best use of human resources (presumably including his own) was the treatment and prevention of AIDS, you would think that he would dedicate his energy to that goal. Or perhaps he would choose a more general task of trying to get world governments to divert more of their negligible international aid dollars to famine relief in sub-Saharan Africa. But that's not what Lomborg does. Rather, he appears to spend most of his time trying to get people to dismiss climate change as too expensive—too low a priority in a world of hard choices. His most recent book, *Cool It: The Skeptical Environmentalist's Guide to Global Warming*, has been an echo chamber best-seller and has created months of new work for the tireless fact-checker Kåre Fog, who has found just as many errors, omissions, and sweeping generalizations in this as in Lomborg's last text. (For example, Fog found nineteen mistakes just within the three pages dealing with polar bears.)

One of Lomborg's most articulate critics is the Silicon Valley software wizard John Mashey, who, on January 8, 2009, posted a devastating critique of Lomborg's prioritizing on a blog called The Way Things Break. In it, Mashey suggests that Lomborg's whole approach is a carefully constructed ruse to avoid paying *anything* toward climate change mitigation. According to what Mashey describes as the Lomborg method, you can avoid almost any spending issue that doesn't suit your political or economic preferences. You begin by proposing a list of alternative priorities that include useful, desirable items that everyone must agree deserve attention—the treatment of AIDS or the provision of food and water to the desperate. Then you make sure that these are items that, for political reasons, will never get funded (foreign aid is a low political priority, especially in difficult economic times). Finally you invoke the false dilemma: you suggest that your audience must accept your prioritization, because if they can't (or won't) pay for items on the top of the list, it would be irresponsible to start thinking about paying for items that are a lower priority.

Tidy though Lomborg's argument may seem when he's making it, it contains more than a few logical fallacies. First of all, if you think about the way you prioritize spending in your own home, there is no question that you would set as the highest priority attending to the dietary requirements and good health of your family. But if your roof sprung a leak, you wouldn't ignore it just because the grocery budget was tight. The second point that Lomborg overlooks is that climate change will make many of his other priority items worse: it particularly threatens the supply of water in the areas of the world where water is already a serious problem, and it poses a huge risk to agriculture across the mostly impoverished equatorial regions.

The beauty of the Lomborg system is that he looks like a social and environmental champion. But even as he is setting up a priority list that stresses attending to "liberal" issues such as AIDS and famine relief, he is busy taking speaking gigs from think tanks that are delighted by his suggestion that climate change mitigation is just too darned expensive.

For the record, Lomborg is a close associate to Danish prime minister Anders Fogh Rasmussen, who led a right-of-centre coalition government from 2001 to 2009. It was Fogh who, in 2002, appointed Lomborg to the newly created Environmental Assessment Institute and who gave Lomborg government support to establish the Copenhagen Consensus Center. In subsequent years, Fogh's government backed away from its aggressive position as the world leader in wind power while at the same time cutting its international aid budget by roughly 25 percent. It had once been the world aid leader, giving 1 percent of its GDP to foreign aid, but it cut that budget consistently, until by the end of 2008, the country had slipped into fourth spot, giving 0.76 percent.[1] It appears that Fogh was happy to pay for Lomborg's advice—especially the part about ignoring climate change—but was equally willing to ignore Lomborg's highest priorities.

"THE ENVIROS WANT the entire store, but I don't think they can fit it in their car."

With that comment, in an interview with Richard Littlemore in February 2009, the loquacious Frank Maisano dismissed the ambitions of an environmental community that, at the time of writing, was still celebrating the recent election of President Barack Obama. Accorded the title and rank of Senior Principal at the law firm of Bracewell & Giuliani (B&G—and yes, that would be Rudy Giuliani), Maisano is not a lawyer. He is listed on the firm's Web site as a "skilled media specialist" and "co-head of the firm's strategic communications practice." In that role Maisano runs interference for lawyers who are trying to help big coal-fired utilities (or big wind-farm developers) site new projects. The five hundred lawyers and support staff at B&G specialize in leading developers through the sometimes byzantine permitting process, and they found that frequently their clients were getting beat-up in the local media. It was often difficult, and always time-consuming, to find local help to "manage" reporters, so B&G decided to set up its own war room, with Maisano in a leading role.

It's a fascinating function, and B&G are innovators in the field. "There are a few [legal] groups that have partnerships with public relations firms," Maisano said in the interview. "But nobody else does this in as detailed and intentional a way as we do."

And Maisano is nothing if not intentional. Leveraging a fifteen-year career on Capitol Hill as a press secretary for Republican congressmen from Michigan and Wisconsin, he knows (and is known in) the corridors of power. From a media perspective, he also has one of the most impressive Rolodex files in the nation. For example, it is impossible to have a conversation while walking with Maisano through the crowds at a Society for Environmental Journalists conference, because nearly every reporter in the room either stops to chat or just shouts, "Hi, Frank!"

It's no wonder. Frank Maisano gives good service. First, he sends out a weekly "energy update" to more than three thousand recipients—mostly to reporters, but also to movers and shakers in industry, politics, and the think tank brigade. The update is a quick, readable wrap-up of interesting stories from the past week and a wide-ranging look at what's coming up in the days and weeks ahead. For reporters on the environment or energy beats, it would be a gift, keeping them current and saving them a considerable amount of leg work. If you don't look too closely, it also seems vaguely evenhanded, talking about green energy projects like major wind farms at the same time as updating people on the latest developments in the coal-fired utility world.

Maisano doesn't stop there. He also sends out instant responses to breaking news, complete with quotes from lawyers who are undeniably experts in the field. He also includes contact information for those experts, so if reporters want more background information, they can get it from B&G.

All of this is good garden-variety public relations. In the PR business, it's our job to make reporters' lives more manageable—and to do everything we can to communicate effectively on behalf of our clients. So it would be hard to criticize what Maisano does or how well he does it. His output is substantive and consistently more forthright than, say, Bjørn Lomborg's twisted arguments. In person, Maisano is gregarious and charming in the hail-fellow-well-met style of a former jock (he was a junior hockey player, who now coaches his own kids and refs twice a week). It's hard not to like the guy.

But it may be dangerous to interpret his bonhomie for sincerity. Maisano's weekly Listserv output shows that he makes a consistent and compelling argument that climate change, though real, is either impossible or just too expensive to fix. For example, in a closing report about the last "UN Framework Convention on Climate Change" conference that Maisano sent out on his Listserv email on December 15, 2008, he complained that international negotiators weren't paying enough attention to economic factors, saying, "If anything, most people seemed to keep the message that no matter the economic conditions, something must be done. *That seems to be an unrealistic approach,* given most of the globe is suffering significant economic woes" (my emphasis).

There's something matter-of-fact in this, something almost self-evident. This is a very difficult time, politically and economically, to start talking about aggressive policy changes to address global warming. But Maisano also either overlooks or misrepresents a point that should be equally obvious. Having just come through fifteen years of robust economic growth, during which we went backwards in our effort to address climate change, he suggests that we can continue our retreat on the basis that an economic downturn makes responsible action unrealistic. Returning to the analogy of someone whose roof has sprung a leak, you cannot dismiss necessary home maintenance just because the budget is tight. You might forego buying a brand-new roof until you get a better job, but you still find a way to apply a workable patch. Besides, the leaky-roof metaphor is probably more appropriate for the ozone hole (another issue entirely). The relevant metaphor for climate change has more to do with heat. We have started a fire that is growing out of control—a fire that has the capacity to render our global home uninhabitable. Dismissing concern about that fire as "unrealistic" is irresponsible in the extreme.

As you might expect from someone advocating for energy development, Maisano also argues consistently that current environmental regulations are sufficient and that resources, once discovered, should be exploited, no holds barred. In a December 22, 2008, email he complained that court challenges had created delays that forced Shell to cancel an exploratory drilling program in Alaska's Beaufort Sea. Maisano says, "These resources can be developed safely and in an environmentally responsible way. In fact, the environmental protections in place in the U.S. are among the most stringent of any in the world." Five items later, in the same email, Maisano celebrates the opportunities that exist for "Brownfield Renewable Energy Developments"—basically the potential to build wind farms on contaminated lands. He writes, "EPA [the U.S. Environmental Protection Agency] estimates that there are about a half million potentially contaminated properties covering approximately 15 million acres across the U.S."

So the U.S., with environmental protections in place that are "among the most stringent of any in the world," has still rendered 15 million acres so toxic that they are not fit for much more than wind farming. This picture of unrestrained and "realistic" progress is chilling if you let it play out in perpetuity.

Still, you have to marvel at Maisano's ability to present half a million contaminated sites as a good thing. And listening to him, you might find yourself nodding in agreement in spite of yourself. In the February 2009 interview with Littlemore he said, "These issues are so complex—there are so many facets—that you have to step back to get a grander picture of what we are really intending to do." That's what he and the team at B&G are trying to do. "We just want to help people find a better balance in how we address climate change, how we provide for our energy needs and how we make sure that people have economic opportunities."

It sounds right, this notion of stepping back, of finding a better balance. But Maisano always seems to hold the picture frame tight enough that all we can see is the economy (and especially the economic opportunity for the companies his firm represents). He never steps back far enough to get a full view of the damage that we are doing to the world and the implications for the billions of people who will never enjoy the economic opportunities of which he now speaks.

Finally, while never quite dipping into the language of denial himself, Maisano maintains his street cred in the denier camp by circulating a steady flow of skeptical reports from places like D.C.–based think tank the George C. Marshall Institute. He also promotes deniers and denier events in his emails: "More than 70 of the world's elite scientists specializing in climate issues will confront the subject of global warming at the second annual International Conference on Climate Change in New York City March 8–10." And he lauds those who have stood up against factual science and common sense in their enthusiasm to present an alternative view: "Another sad note this week as we heard about the passing of the great writer Michael Crichton."

For a last word on Maisano, however, it's hard to improve upon this comment from Associated Press science writer Seth Borenstein: "Frank is the good cop to Marc Morano's bad cop."

It's an astute assessment of the role both men play. Morano's willingness to be totally outrageous in his promotion of denial opens up considerable middle ground, which means that Maisano looks like a pillar of reason by comparison. After dealing for fifteen years with people who refused even to acknowledge that the science is real and the situation serious, it's a relief, at first, for environmental journalists to talk to someone like Maisano, who, like Lomborg, begins with a smile and a willing admission that the scientific explanation of human-induced global warming is undeniable. Yet both men come to the same conclusion that Morano has been promoting all along: they suggest that action is unnecessary, unaffordable, or unrealistic. And Maisano, picking up the bar tab in a room full of environmental reporters, offers condolences while gathering up everyone's business card—the easier to ensure that the next time they're writing a story, they'll be well-served with a big helping of B&G spin.

WHEN IT COMES to staking out positions and shifting the middle ground, industry-funded strategists seem to have seized and kept the strategic initiative. Any time anyone on the science side makes even the smallest overstatement, they immediately face the full resources of a think tank echo chamber attack. And because conscientious scientists are so quick to recognize and acknowledge when something is not exactly correct, the attackers have won many apologies, corrections, or reinterpretations, which they have used to argue that all of climate science is frail and uncertain.

At the same time, the more exuberant deniers have often said things that were flat-out wrong, and then have refused to acknowledge or apologize for their misrepresentations. In response the environmental community—lacking both the resources and the sense of common purpose more typical of the antiscience crowd—has been ineffective in launching countercriticism.

The February 15, 2009, George Will column in the *Washington Post* that I mentioned in Chapter 8 was a disappointing example. Will dismissed concern about climate change partly on the basis that global sea ice, frequently reported to have declined dramatically over the past several years, was actually equal in extent to what it had been when monitoring began in 1979.

This would have been compelling if it were true. But, thanks in part to the resulting controversy, the National Snow and Ice Data Center in Boulder, Colorado, issued a press release clarifying the situation a few days later, stating, "We do not know where George Will is getting his information, but our data show that on February 15, 1979, global sea ice area was 16.79 million sq. km and on February 15, 2009, global sea ice area was 15.45 million sq. km. Therefore, global sea ice levels are 1.34 million sq. km less in February 2009 than in February 1979. This decrease in sea ice area is roughly equal to the area of Texas, California, and Oklahoma combined."

Yet Will stuck to his guns, and the *Washington Post* insisted that the piece had been adequately fact-checked. *Post* editor Fred Hiatt defended Will's right to his opinion and offered to open his pages to further debate. And the *Post*'s ombudsman, Andrew Alexander, followed up on February 28 in a column headlined, "The Heat from Global Warming," saying that the paper had every right to "present a mix of respected and informed viewpoints."

That would be true—if the information were true. The fact that it clearly was *not* true should have impressed the editors at the *Post* at least to the point of eliciting a correction. People weren't complaining about George Will's opinions. They weren't trying to engage him in debate. They were complaining that he had misrepresented the facts—the first time perhaps inadvertently, and the second time through stubborn reassertion. And when you reassert something you know to be false, that could easily be called lying, not debating.

Into the midst of this fracas waded the *New York Times*'s Andrew Revkin, probably the best-read and most influential science and environment reporter in the country. Revkin is something of a hero in the science community for his hard work and his felicity to the facts. Between his stories in the *Times* and his additional writings on his popular blog DotEarth.com, he has been a leading voice for reason and accuracy on the climate file. But Revkin still hails from that world where balance is sometimes honored over accuracy. In a February 24, 2009, *Times* story, "In Climate Debate, Exaggeration Is a Pitfall," Revkin reacted to the George Will debacle by comparing it to a recent instance in which former vice president Al Gore had been forced to remove a slide from his usual presentation because the scientific authors of the information said that Gore had overstated the impact of climate change on weather-related catastrophes. Revkin concluded, "In the effort to shape the public's views on global climate change, hyperbole is an ever-present temptation on all sides of the debate."

Under the circumstances Gore might reasonably have been accused of hyperbole—exaggeration or overstatement. But that's different from lying. Gore also removed the offending slide from his presentation immediately. But Will simply repeated the misinformation, and the editors of his newspaper stood up in defense of his right to do so. Then Andy Revkin—again, probably the best and most credible climate change reporter in North America—offered this up as an indication that each side in the debate is sometimes guilty of not playing strictly by the rules.

Thus does the "middle ground" shift into unreality. The "reasonable" voices, of which Revkin must certainly be one, wind up giving assent to a scientific controversy that in fairness, and on the basis of the best and most accurate science, does not exist. With people like Benny Peiser and Marc Morano staking out the unreasonable fringe, with Fox News quoting energy lobbyists such as Steven Milloy as "experts," people like Bjørn Lomborg and Frank Maisano start looking like the most reasonable characters around. They establish the middle ground, and they say, almost regretfully, that serious though climate change may be, it's just too—what's that word?—inconvenient to fix it just now. And the media look at the range of opinions and say, yes, there are exaggerations on all sides, suggesting that we should all be careful to stick to the muddied middle path.

An April 7, 2009, *Reuters News Service* story, "EU: Earth Warming Faster," gave a chilling prediction for where that middle path leads. Reuters had chosen the eleven leading European scientists—nine of them lead authors of the last report of the IPCC—and asked for an update. They said warming was accelerating beyond what had been anticipated in the last report and that it was even more certain that humans were to blame. Worse, they said it was "unlikely"—which to them means there is less than a one-third chance—that we will limit the global temperature increase to four degrees Fahrenheit. And that increase, they say, is the trigger point beyond which climate change becomes truly dangerous.

It seems that there are no safe compromises to be made in dealing with climate change. Denying it was wrong. Delaying action is dangerous. People who say otherwise should, at some point in the very near future, have to stand accountable for their recklessness.

[*eleven*]

SLAPP SCIENCE

Using courts and cash to silence critics of climate confusion

A Strategic Lawsuit Against Public Participation (SLAPP) is a lawsuit intended to intimidate and silence critics by burdening them with the cost of a legal defense until they abandon their criticism or opposition. Winning the lawsuit is not necessarily the intent of the person filing the SLAPP. The plaintiff's goals are accomplished if the defendant succumbs to fear, intimidation, mounting legal costs, or simple exhaustion and abandons the criticism. A SLAPP may also intimidate others from participating in the debate.

WIKIPEDIA, APRIL 30, 2009

It's a tricky business, accusing someone of filing a lawsuit merely to silence their critics. By saying so, you must necessarily suggest motive on the part of the plaintiff, which can be difficult or impossible to prove. So for the record, I have no direct evidence that the suits I am about to explore were SLAPPs. In the first case, plaintiff Dr. S. Fred Singer prevailed, winning an apology from his target, Justin Lancaster. In the second case the plaintiff, Tim Ball, bailed without comment as soon as the target put up a fight. And in the third case, plaintiff Stuart Dimmock lost, but by cleverly spinning the result—and by recruiting the assistance of ideologically primed journalists—his supporters were able to turn defeat into a bizarre kind of victory.

The first suit keyed off the reputation of Roger Revelle, a giant of science whom you may remember from Chapter 2 as one of the early voices raising alarm about the potential danger of global warming. Revelle is also an icon of climate science, in part because Al Gore speaks of him so fondly. Revelle taught Gore at Harvard during the 1960s, explaining even then the complications that might evolve if humans continued to pour greenhouse gases into the atmosphere without regard for the consequences. It's clear from the recent activities of the Nobel Peace Prize–winning former vice president that Revelle had the capacity to engage and inspire. Gore never forgot the lesson and has been one of the most influential advocates in the world for action on global warming.

But Revelle had also caught the attention of people who were less enthusiastic about acknowledging the threat of unrestrained burning of fossil fuels. One of those people was Dr. Siegfried Frederick Singer, who as early as 1990 was arguing for inaction. In fact, in an article written for the journal *Environmental Science & Technology* in 1990, Singer concluded that for economic reasons, we should be wary of overreacting to climate change: "Drastic, precipitous, and especially, unilateral steps to delay the putative greenhouse impacts can cost jobs and prosperity without being effective... It would be prudent to complete the ongoing and recently expanded research so that we will know what we are doing before we act. 'Look before you leap' may still be good advice."

That may have sounded like cautious wisdom in 1990. It most certainly sounded like the words of a man who was in no hurry to disrupt the status quo. But Singer wasn't content to have those words printed only under his own name. It appears that he wanted to be able to attribute those thoughts to someone else—say, for example, someone who was widely trusted in the world as a credible commentator on climate change. Say, Roger Revelle. So Singer convinced Revelle to sign on as a coauthor for a "new paper" to be published in *Cosmos*, the peer-reviewed journal of the prestigious Cosmos Club. A third coauthor was the late Chauncey Starr, founder of the Electric Power Research Institute and a founding "science advisor" on Singer's Science and Environmental Policy Project.

In 1991, at the time this project was unfolding, Revelle was in his eighty-second and, it would turn out, last year. He had already suffered a serious heart attack and was in failing health—unable, according to his students and staff, to pay attention for more than fifteen or twenty minutes. Singer brought Revelle galley proofs of the paper and stayed with him for three hours, purportedly reviewing the finer points of the science. Then he published an article that concluded with: "Drastic, precipitous—and, especially, unilateral—steps to delay the putative greenhouse impacts can cost jobs and prosperity and increase the human costs of global poverty, without being effective... It would be prudent to complete the ongoing and recently expanded research so that we will know what we are doing before we act. 'Look before you leap' may still be good advice." This, of course, seems more than coincidentally similar to Singer's own conclusion—a similarity that prevails through most of the article.

Justin Lancaster, who was working with Revelle as a graduate student at the time, was offended by what he saw as Singer's blatant manipulation—his outright bullying—of Revelle. In an account of the incident posted on the Environment Science & Technology Web site, Lancaster wrote, "Revelle was hoodwinked, in my view. Perhaps more severe terms are deserved. My personal conversation with Roger shortly after the publication of the *Cosmos* article gave me the very strong sense that he was intensely embarrassed that his name was associated. He seemed noticeably relieved when we agreed together that perhaps the readership of the Cosmos Club journal would be small and limited. Little did either of us anticipate at that moment what was really going on and what would follow." What followed the article's publication was an early application of the echo chamber. As the world's great powers were gathering in Rio de Janeiro for the Earth Summit in 1992, Singer and company began spouting everywhere the news that the great Roger Revelle—grandfather of the greenhouse effect—had capitulated. Lancaster lost his temper and said publicly that Singer had duped Revelle. Ill-advisedly, Lancaster also suggested that Singer's actions were unethical.

The ensuing lawsuit hit hard. Lancaster says now that he suddenly found himself sitting in the offices of some of the most expensive lawyers to be found anywhere in the northeastern United States, staring into the black hole that had once been his future. Here he was, a recently graduated Ph.D. with no job and, thanks to this controversy, no offers. Lancaster's own lawyers looked at the evidence and said that Lancaster had a very good case to make before the courts. The similarities between the original Singer article and the one purportedly coauthored with Revelle go well beyond the conclusions quoted here: a compelling case could be made that Singer was the author on both occasions.

But the lawyers also told Lancaster that using the word "unethical" is always dangerous and usually expensive. They said that even if Lancaster decided to pursue the case and won on the main point, the challenge to Singer's ethics would be difficult to defend. And even in victory Lancaster would still be left with a legal bill so huge that he and his young family might never recover.

So Lancaster rolled over, retracted his charge, and walked

away. And a decade later Singer was still crowing. He penned a chapter ("The Revelle-Gore Story: Attempted Political Suppression of Science") in 2002 in a Hoover Institution book titled *Politicizing Science: The Alchemy of Policymaking,* and he has continued to quote Revelle's capitulation and to charge that Lancaster had only complained because he was doing Al Gore's bidding.

Unable to find work as an academic because of the embarrassment of the whole affair, Justin Lancaster moved off into the private sector, becoming successful enough that he now has the resources to stand up for principle. He has "retracted his retraction" and now says unequivocally that Singer suckered Revelle—taking advantage of a kind and intelligent but aging intellectual. Although Lancaster's own version of the events has been on the Internet for almost three years, he has heard no more from Dr. S. Fred Singer.

Even so, Lancaster's earlier capitulation had made the whole scientific community nervous. It can cost tens of thousands of dollars to defend yourself in a libel trial, and even if you win, you never recover all of your investment or any of your time. Facing an industry-backed player like Singer, many other scientists have likely chosen the prudent course and stayed out of the debate.

As an aside—on the question of whether Singer is, in fact, industry backed—it's worth mentioning Singer's connection to the junk science crowd. To his credit, there is no doubt that Singer was once widely considered to be a distinguished physicist. In the early 1960s he was a leader in developing Earth-observation satellites. He was a special advisor to U.S. president Dwight D. Eisenhower and a founder and the first director of the U.S. National Weather Bureau's Satellite Service Center. For more than twenty years, however, he has been closely associated with the think tanks and lobby groups that are fully engaged in supporting the kinds of industries that might attract negative public attention. For example, Singer has argued that the use of DDT is a good thing and should be restored. When 6,500 American physicists condemned Ronald Reagan's Strategic Defense Initiative as impractical, Singer joined a very exclusive club of scientific cheerleaders. And, as reported in the Chapter 8 discussion about petitions disputing the science of climate change, Singer has long been a leader in the campaign of confusion. His Web site, the Science and Environmental Policy Project, is a catchall for "scientific" information favorable to the industries with which he associates.

But Singer doesn't like criticism, especially if you report that he has served as a paid defender of the tobacco industry. When the DeSmogBlog suggested such a connection in a post on June 16, 2006, Singer responded on June 18, 2006, with a threat to sue and a demand that we post a retraction, saying, "Dr. Singer and SEPP (Science & Environmental Policy Project) have no connection whatsoever with the tobacco industry, now or in the past." But the record suggests otherwise. It appears from a series of memos filed on the Web site tobaccodocuments.org that Singer was engaged, at least once, to write a "research" paper challenging the health effects of secondhand smoke. For example, there is an internal memo from the Alexis de Tocqueville Institution commending Dr. Singer as "the man [who] would handle the EPA/ETS [environmental tobacco smoke] . . . work." Commenting on Singer, the March 1994 memo's author says, "Very impressive resume—I think the project is worth the [US]$20 K we discussed." In a December 2004 letter to Walter Woodson, vice president and director of public affairs at the Tobacco Institute, de Tocqueville executive director Cesar Conda thanks the Tobacco Institute for the US$20,000.

The site also contains a copy of a draft report, "The EPA and the Science of Environmental Tobacco Smoke," on which Singer is listed as the lead author. And there is an August 1994 memo from Tobacco Institute president Samuel D. Chilcote

Jr. celebrating the press conference at which the institute and the de Tocqueville Institution jointly released the report, with Singer and Jeffreys in attendance. The paper itself was a bit of a stretch as a scientific analysis. Presented as a "peer-reviewed document," it turned out that most of the nineteen reviewers, like Singer, had little or no expertise in the science of smoke. Ten of the reviewers were economists. Two were fellows at another libertarian public policy institute (the Hoover Institution), and one was a mineralogist. Singer's fairly political conclusion in this "scientific" paper was as follows: "I can't prove that ETS [environmental or secondhand tobacco smoke] is not a risk of lung cancer, but the EPA can't prove that it is." He seems to be both promoting and taking comfort from uncertainty.

On the strength of this evidence, I declined to apologize, and, though Singer's lawyer sent two more threatening letters, they never followed through on a suit.

This was not the first occasion that Singer had been reported to have taken money from industry only to deny it after the fact. For example, on February 12, 2001, Singer wrote to the *Washington Post* in a letter to the editor that the *Post* headlined "My Salad Days," denouncing suggestions that he had accepted money to lobby on behalf of the fossil fuel industry. "As for full disclosure," he said, "my résumé clearly states that I consulted for several oil companies on the subject of oil pricing, some 20 years ago, after publishing a monograph on the subject. My connection to oil during the past decade is as a Wesson Fellow at the Hoover Institution; the Wesson money derives from salad oil." Yet Ross Gelbspan has reported on HeatIsOnline.org that ExxonMobil's own Web site listed Singer as a recipient of US$10,000 in direct funding in 1998 and as a participant in an event that year to which ExxonMobil contributed US$65,000. Asked since about those payments, Singer has taken the amnesia defense: he says he gets so many checks that he can't remember who they're all from.

ON APRIL 19, 2006, the self-appointed Canadian climate change expert Dr. Timothy Ball published an opinion column in the *Calgary Herald,* the newspaper of record in Canada's oil capital (often referred to as "Houston North"). The column, headlined "Aussies' Suzuki Heavier on Rhetoric than on Science," was a churlish attack on the Australian academic Tim Flannery, who was crossing Canada on a promotional tour for his best-selling climate change book, *The Weather Makers.* Ball derided Flannery as someone with "no professional credentials in the field" who "blunders regularly." He alleged that Australian scientists had "debunked" Flannery's book, and he stated flatly that "there is no more reason to believe Flannery today than there was" to believe Lowell Ponte, a scientific outlier who warned of global cooling in the 1870s.

All this was fairly typical invective from Ball, who had garnered a national reputation as a letter and opinion-page writer and an exuberant speaker on the climate change denial circuit. Also typical was this biographical note, added at the bottom of his column: "Tim Ball is a Victoria-based environmental consultant. He was the first climatology Ph.D. in Canada and worked as a professor of climatology at the University of Winnipeg for 28 years."

Four days later the *Herald* printed a letter to the editor from an Alberta academic named Dan Johnson under the headline "Vital Statistics." It began, "Whatever one may feel about Tim Ball's denial of climate-change science, newspapers ought to report factual summaries of authors' credentials." Johnson, a professor of environmental science and a Canada Research Chair at the University of Lethbridge, went on to name several esteemed Canadians who received climatology Ph.D.s before Ball, and he added that "it is important to recognize them and their research." Johnson also noted that Ball was *not* a climatology professor at the University of Winnipeg for twenty-eight years (Ball later confirmed that the correct number was eight).

Johnson then concluded that Ball's work "does not show any evidence of research regarding climate and atmosphere and the few papers he has published concern other matters. There are great gains to be made in science from conjectures and refutations, but sometimes denial is nothing more than denial."

There followed a series of threats and arguments between Ball and the *Calgary Herald,* culminating on September 1, 2006, with a thumping great lawsuit, Tim Ball railing against the perfidious conduct of the newspaper and castigating Dan Johnson for impugning Ball's character and reputation. Ball demanded C$325,000 in general and punitive damages, plus costs and damages for lost income in an amount to be set at trial.

Given the weight of evidence supporting Johnson's position, this might have been passed off as a minor nuisance, except for this: defending yourself against such a lawsuit can be prohibitively expensive. At the time the suit arrived, Dan Johnson's new wife, Julie, had just delivered twins, one of whom had a congenital malformation of his foot that required a long series of surgeries in the first six months and a painful and restricting leg brace. Julie was also trying to complete her own Ph.D., leaving the family to wrestle with enough stress—even aside from the threat of a $325,000 judgment.

Yet everything that Johnson had said was easily subject to the libel defense of truth. There really was a long parade of highly respected climatology Ph.D.'s before Ball came along, and Ball had not been a professor for twenty-eight years. In fact, twenty-eight years before his retirement in 1996, Ball was still pursuing his bachelor's degree at another university. Yet the Friends of Science front man had been advertising himself hither and yon as the first climatology Ph.D. in the country and had said in public and on Web site biographies that became part of Johnson's statement of defense that he had been "for 32 years a Professor of Climatology at the University of Winnipeg." Ball even wrote a letter to then Canadian prime minister Paul Martin in which he said, "I was one of the first climatology Ph.D.s *in the world*" (my emphasis). He signed that letter, "Dr. Tim Ball, Environmental Consultant, Victoria, British Columbia, 28 Years Professor of Climatology at the University of Winnipeg." (And no, the "32 years" reference was not a typographical error, at least not here. Ball used different years in different biographies.)

You would think someone who had been so casual in remembering the details of his career would be reticent to pick a fight over those details. But perhaps Tim Ball thought that Dan Johnson would, like Justin Lancaster, look at the cost of defending himself and choose instead to apologize and retract his criticism. There was certainly pressure to do so. The administration at the University of Lethbridge made it clear that they would offer no assistance with Johnson's defense, and his colleagues were silent while he took the heat.

But Dan Johnson chose to stand and fight. Ultimately even the *Calgary Herald* did the right thing, commissioning a statement of defense that was nearly as devastating as Johnson's own. One of the best parts was the *Herald's* answer to whether Johnson had damaged Ball's reputation: "The Plaintiff [Ball] never had a reputation in the scientific community as a noted climatologist and authority on global warming . . . The Plaintiff is viewed as a paid promoter of the agenda of the oil and gas industry, rather than as a practicing scientist." It is unfortunate that this strident criticism was never reported in the news pages of the *Calgary Herald*.

Under the weight of the evidence Ball capitulated the following year, withdrawing his suit quietly and without comment. But there was no justice to be had for Dan Johnson. The *Herald*, which had been righteous in its defense about having a "moral or social duty to present both sides of the global warming debate to its readership," has forgotten that duty, failing even to report that Ball had withdrawn his suit and that Dan Johnson was exonerated.

This was made worse by a "clarification" that the *Herald* had run at Ball's request on August 20, 2006, a week before Ball launched his suit. Against Dan Johnson's statement that Ball had not published on climate change, the *Herald* wrote, "According to Ball's curriculum vitae, he has conducted research on climate and has published 51 papers—32 directly related to climate and atmosphere." (That's what they get for taking Ball's CV at face value. A search of the ISI Web of Knowledge, the most exhaustive database available for recording peer-reviewed scientific papers, showed at the time that Johnson's statement of defense was filed that Ball had a lifetime output of just four peer-reviewed papers, none touching even distantly on atmospheric science.)

If you go to the FPinfomart Web site, on which the *Calgary Herald*'s stories are available (at a charge) for public viewing, you will find (as of April 30, 2009) that this erroneous correction is still attached to Dan Johnson's letter, but nowhere will you find the resolution to the case, setting straight the facts that Johnson was correct and Ball, indefensibly, in the wrong. So in a fundamentally important way, Ball enjoyed a kind of victory. Dan Johnson wound up fighting for *his* reputation and, because of the *Herald*'s neglect in failing to set the record straight, appears to be the one whose reputation is still being—falsely—maligned. Ball, who should have been humiliated by the facts, continues to write antiscience diatribes and give lectures, relatively safe in Canwest Global's cone of silence.

THE NEXT LEGAL target was Al Gore. In fact, the think tanks' favorite target *always* seems to be Al Gore. When the Heartland Institute props up Christopher Monckton as a debating champion, the opponent they most dearly want to face is Al Gore. When Fred Singer suckers Roger Revelle and then sues Justin

Lancaster, the person he really wants to discredit is Al Gore. Even if the Nobel committee had failed to honor the former vice president for his tireless efforts at climate science education, they might have accorded him a Peace Prize just for having survived the last ten years without getting into a physical fight with any of his legions of think tank detractors.

In the case at hand, Stuart Dimmock v The [U.K.] Secretary of State for Education and Skills, the U.K. *Telegraph* reported in September 2007 in "Lorry Driver in Challenge to Gore Film" that Dimmock, a truck driver and "school governor from Kent," was taking the government to court over a decision to promote the showing of Al Gore's film, *An Inconvenient Truth,* in U.K. high schools. The earnest Dimmock told the *Telegraph,* "I care about the environment as much as the next man. However, I am determined to prevent my [two teenaged] children from being subjected to political spin in the classroom." There were no early reports as to who might be financing this putative act of sacrifice on Dimmock's part.

The stated purpose of the court case was twofold. First, Dimmock asked the court to ban *An Inconvenient Truth* from U.K. schools on the basis that it was scientifically flawed and politically motivated. Failing that, Dimmock's lawyers argued (in the words of Justice Michael Burton) "that, if the political issues, as per the content of [the film], are to be brought to the attention of pupils, then there must be an equivalent and equal presentation of counter-balancing views."

Which is to say, schools must be forced to match *An Inconvenient Truth* with a denier film such as *The Great Global Warming Swindle*—which was a swindle, indeed. While *An Inconvenient Truth* has attracted a huge amount of very critical attention from Exxon-funded think tanks and the junk scientists they employ, the *Swindle,* composed by a controversial U.K. documentary filmmaker named Sean Durkin, has been criticized instead by the scientific community, including, and perhaps especially, scientists who were actually conned into participating in the film, only to have Durkin misuse or misrepresent their work.[1]

The court case itself moved quickly from science to theatre, with the plaintiffs calling expert witnesses such as Bob Carter, one of Australia's most prominent deniers, to criticize several points in Gore's film. Neither was it, as some of the later news coverage implied, a judicial testing of all the theories of climate change and global warming. Rather, it was a very specific argument over whether *An Inconvenient Truth* could be presented in U.K. schools—at all, and on its own. In the end the judge rejected both of Dimmock's demands. Justice Burton first found no reason to ban the film from the classroom, saying that it is "substantially founded upon scientific research and fact." He granted that Gore is "a talented politician and communicator, [trying] to make a political statement and to support a political programme [the mitigation of climate change]," but he said this about the film in general:

> The Film advances four main scientific hypotheses, each of which is very well supported by research published in respected, peer-reviewed journals and accords with the latest conclusions of the IPCC:
>
> 1. global average temperatures have been rising significantly over the past half century and are likely to continue to rise ("climate change");
> 2. climate change is mainly attributable to man-made emissions of carbon dioxide, methane and nitrous oxide ("greenhouse gases");
> 3. climate change will, if unchecked, have significant adverse effects on the world and its populations; and
> 4. there are measures which individuals and governments can take which will help to reduce climate change or mitigate its effects.

Justice Burton found that these propositions, as put forth by the government's lawyer, Martin, are supported by a vast quantity of research published in peer-reviewed journals worldwide and by the great majority of the world's climate scientists.

As to the argument that students should have to endure a contrarian film such as the Durkin *Swindle,* Justice Burton said, "There is nothing to prevent (to take an extreme case) there being a strong preference for a theory—if it were a political one—that the moon is *not* made out of green cheese, and hence a minimal, but dispassionate, reference to the alternative theory. The balanced approach does not involve equality." The language is tortured, but the message seems clear: letting Sean Durkin's film into the school system would be tantamount to conscientiously exposing children to the view that the moon really is made of green cheese.

So the Dimmock case failed, right? Well, he lost on both points that he had originally pressed. Justice Burton approved the continued showing of *An Inconvenient Truth* in schools, albeit with a more specific set of accompanying notes, and he rejected out of hand Dimmock's request to stick Durkin's *Swindle* into the schoolchildren's package.

But the supporters and financiers of Dimmock's case created a stunning success from this apparent defeat. In contemplating the evidence, Justice Burton had accepted that there were nine instances in Gore's film in which you could argue that he had erred on the science or overstated what was, at the time, the global scientific consensus. The judge advised that these potential overstatements be pointed out to students who are shown the video during school.

Armed with that part of the verdict, rather than slinking out in defeat, Dimmock and company left the courtroom with fist-pumping enthusiasm, claiming that the judge had condemned Gore's film. The international media completely swallowed their spin, as these headlines that followed reveal:

"Judge Attacks Nine Errors in Al Gore's 'Alarmist' Climate Change Film"
—*The Evening Standard,* October 11, 2007

"Al Gore's Inconvenient Judgment"
—*The Times Online,* October 11, 2007

"Gore Climate Film's 'Nine Errors'"
—BBC *News,* October 11, 2007

"An Inconvenient Verdict for Al Gore"
—ABC *News,* October 12, 2007

In the most outrageous example, the Canadian newspaper the *National Post* carried two and a half pages of coverage on October 11, 2007, beginning with a headline that read, "U.K. judge rules Gore film 'exaggerated,'" and nowhere in the whole paper did the *Post* mention that Justice Burton had categorically dismissed Dimmock's principal complaints.

There it was, in virtually every newspaper and broadcast in the English-speaking world. Rather than being presented as an exoneration of *An Inconvenient Truth,* the court case was made into a condemnation—a humiliation for Gore and a repudiation of climate science.

Like the early Edward Bernays ploys from Chapter 3, this stands as another example that you can imagine a public relations class studying in the future—the professor pointing out the tactical skill of the Dimmock group at winning a huge victory in the court of public opinion, regardless of having been rebuffed in a court of law. What you can't imagine is that anyone in the Dimmock party ever stopped to ask themselves whether what they were doing was right.

And who was in the "Dimmock party" anyway? Was this

really a lone father, standing up for the rights of his 11- and 14-year-old sons? Were it not for vanity, we might never have known. But five months after the case was decided, Christopher Monckton turned up on March 4, 2008, on the *Glenn Beck Program* on Fox News, taking credit for the whole affair. Monckton told Beck that after viewing *An Inconvenient Truth* "with mounting horror," he called an old friend who he thought might foot the bill "to fight back against this tide of unscientific freedom-destroying nonsense."

The *Guardian* newspaper had reported on October 14, 2007, in "Revealed: the Man Behind the Court Attack on Gore Film" that Monckton's old friend was Robert Durward, the gravel-pit magnate who had started the U.K. Scientific Alliance and who has been the chief benefactor to the New Party, a libertarian splinter party for which Monckton wrote the original manifesto and Dimmock was an early, though unsuccessful, candidate. When Monckton looks at *An Inconvenient Truth* and says, it's all politics, that is apparently because, for him, it's all politics.

Monckton's other purpose on the Beck show (other than to argue that DDT is actually a wonder chemical that should be sprayed liberally around the developing world) was to try to rustle up funding for an American court challenge to the showing of *An Inconvenient Truth.* Unsatisfied with the facts of the actual judgment, Monckton also left Beck with the impression that the court had commanded teachers to show Durkin's *Swindle* in schools along with the Gore documentary:

> GLENN: Okay. So you told me last night—because I said to you, I have no problem with Al Gore's movie being shown in school, as long as the *Great Global Swindle* or something like that is also shown side by side. Show both sides. Teach both sides. You said that the same thing that happened over in

> Great Britain can be done here, but it requires about $2 million to get it done.
>
> MONCKTON: This is absolutely right. So if any of your listeners out there have got a spare $2 million and you would be willing to take Al Gore on in court, then get in touch with Glenn Beck and he'll get in touch with me...

No takers had emerged at the time of writing. But really, why would you invest in a losing court case when your team has already wrung all the public relations benefits imaginable out of the last round of defeat? On this issue, against someone of Al Gore's reputation and character, it's clearly better to declare victory—and stay in your cave.

[*twelve*]

MANIPULATED MEDIA

An industry overwhelmed in the age of information

I have considered, if only briefly, the state of climate science. As Naomi Oreskes showed in her *Science* survey of peer-reviewed literature, no serious scientific debate exists over the fact and origins of global warming. Despite Benny Peiser's best efforts, no one has been able to refute that finding. On the contrary, the science academies in every major nation in the world have all stepped forward, issuing statements affirming that climate change is a potential global crisis and that humans are both the principal cause and the only hope for a solution. The IPCC, while acknowledging that in science there must always be room for error, has said that there is a greater than 90 percent chance that our spaceship is going to crash if we don't change course. The signatories to the UN Framework Convention on Climate Change—nearly two hundred countries and organizations from around the world, ranging from Albania and Argentina to the United States and Zimbabwe—have agreed that in the absence of an option to climb aboard a different spaceship, we should be fixing the one we have—or at least we should stop wrecking it quite so quickly. That should meet anybody's definition of consensus, while accommodating the inevitable pockets of disagreement that are accepted within the definition. We have a problem (global warming), a diagnosis (the human generation of greenhouse gases), and a suite of proposed solutions (ranging from obvious money-savers such as energy conservation to innovative challenges such as the development of sustainable energy alternatives).

It seems so obvious. Or perhaps not. If you ask Americans whether the climate is changing, just over 70 percent of them will say yes. (Pew Center research from May 2008 put the exact number at 71 percent, while an ecoAmerica poll from October 2008 set the total at 73 percent.) But if you ask how many believe humans are responsible, the number drops below 50 percent.

From a political perspective, that is not enough. No one is going to get elected by promising to solve a problem that half the people don't even acknowledge. And solving climate change is not going to be easy. It will demand major changes in the way we consume energy, shape our cities, and source our food. It will demand an unprecedented international act of social and political will. Such changes cannot be wrought with minority support, even with a newly elected U.S. president Barack Obama heralding a climate of change.

Looking more deeply at the nature of public understanding, you can also see some stunning partisan divisions in how people comprehend climate science. It is an article of faith among some political independents in the United States that there is no real difference between Democrats and Republicans, but on the issue of climate change that couldn't be further from the truth. According to the Pew poll cited above, 84 percent of Democrats say the Earth is warming, compared to just 49 percent of Republicans. And in 2008 the Republican numbers were going down: they had dropped from a total of 62 percent only a year earlier. Asked whether humans are responsible for that warming, 58 percent of Democrats said yes; just 27 percent of Republicans agreed.

But here's where the numbers get really surprising. At least, these are the numbers that most surprised me. If you look among highly educated partisans, the division gets deeper. Among Democrats with a college education, Pew found that 75 percent believe humans are responsible for global warming. Among Republicans: 19 percent.

The only way to explain that split is through the way these groups consume media and the nature and types of media they consume. It's generally safe to assume that people with more education will be better informed. On this issue, however, Democrats seem to be better informed about the science, while Republicans are better informed about the controversy.

You can blame this on a whole host of factors: the complexity of the issue, the attention span of the modern media consumer, the consolidation of major media, the splintering of media markets through the Internet, and the think tank campaign to confuse. But at the end of the day a huge amount of blame has to be laid at the feet of what in the Internet age is known as "mainstream media." Whether through inadvertence, understaffing, or, in certain outlets, an actual intent to misinform, many major media outlets in North America and, to a lesser extent, in Europe have failed to inform their readers and listeners about what is surely the most important and dangerous environmental issue in the history of humankind.

PERHAPS WE SHOULD look first at how we consume information. There was a great feature in the *New Yorker* in December 2007, in which author Caleb Crain chronicled the decline of reading and the implications of returning to an oral culture—one where we depend less on what we read than on what we hear. In the article, titled "Twilight of the Books," Crain documented the decline in book sales and library retrievals and the relative collapse in the reading of newspapers, including online editions. People have grown accustomed, instead, to getting their news from television and radio, two sources that struggle with complexity and that resist careful criticism.

The complexity issue is a real problem for global warming, in part because there are no easy analogies to explain the science: you have to sit down and work through the details. Past scientific issues have not created as much of a challenge. With the ozone hole, for example, people understood the issue metaphorically, if not scientifically: there is a protective coating around the Earth called the ozone layer, a kind of a roof. And because of human activity, we were tearing a hole in that roof. It was so clear a concept that people have even tried to use it to explain climate change. When the David Suzuki Foundation commissioned a national study in 2006 on Canadian understanding and attitudes toward climate change, more people blamed global warming on the ozone hole than on any other factor. Clearly the media—print or broadcast—have not succeeded in transmitting even the most rudimentary explanation of the actual cause of climate change.

When it comes to judging and criticizing sources of information, TV and radio also present a problem for listeners. Crain puts it this way: "It is easy to notice inconsistencies in two written accounts placed side by side. With text, it is even easy to keep track of differing levels of authority behind different pieces of information. The trust that a reader grants to the *New York Times*, for example, may vary sentence by sentence. A comparison of two video reports, on the other hand, is cumbersome. Forced to choose between conflicting stories on television, the viewer falls back on hunches, or on what he believed before he started watching." Thus, while reading offers the opportunity for sober second thought, TV and radio give you only enough time to pick out bits of information that reconfirm your existing biases. Crain also argues that it's harder for radio and television to transmit information that contradicts what you already believe, saying, "It can be amusing to read a magazine whose principles you despise, but it is almost unbearable to watch such a television show."

Democrats forced to listen to Rush Limbaugh or conservative Republicans trapped in an Al Gore lecture on climate change are likely to agree with that statement. The nearly infinite number of media choices now available through the Internet and cable TV have contributed yet further to our ability to become extremely well-versed in arguments that reinforce our political, economic, social, and scientific assumptions. The well-read American business executive, working daily through the *Wall Street Journal* and biweekly through the *National Review* and keeping constantly tuned to Bill O'Reilly and Glenn Beck on Fox News, will be barraged with stories that suggest a scientific controversy about climate change. Every new article that muddies the waters, no matter how obscure, finds its way into the pages of these publications or onto the Fox airwaves. Each winter snowstorm or Florida frost is presented derisively as evidence that the globe is not warming quickly enough. This steady diet of doubt can't help but influence people who are working diligently to stay well-informed.

But the failure of journalism is not merely a matter of partisan news outlets picking out a selection of stories that tend to reaffirm an established audience bias. Even that would be defensible if they maintained a reasonable standard of accuracy and made an effort to weed out obvious manipulation. But in the worst cases networks like Fox and newspapers like Canada's *National Post* throw caution (and sometimes good practice) to the wind, broadcasting or publishing the works of Steven Milloy and Fred Singer without mentioning to viewers or readers that these people are taking money from the industries they are trying to defend.

Milloy may be the most obvious example. He has spent his entire career in public relations and lobbying, taking money from companies that include Exxon, Philip Morris, the Edison Electric Institute, the International Food Additives Council, and Monsanto in return for his work declaring environmental concerns to be "junk science."[1] Smoked out of his position as the executive director of the Philip Morris–created TASSC and revealed as one of the authors of the American Petroleum Institute's plan to sow doubt and confusion about climate change, Milloy has continued to lobby against environmental regulations of all kinds. His Web site, JunkScience.com, is a running attack on environmental issues and the people who promote them.

Yet when Fox News brings Steven Milloy into the studio, they somehow never get around to mentioning that he is an industry lobbyist. They call him a "junk science expert"—which is true in its way, but is in no way helpful in letting viewers identify him for what he is: not an impartial journalist committed to a balanced or even accurate account, but a registered corporate lobbyist who has been paid to spin the issue to the advantage of his clients. It is an offense against the principle of transparency that Fox doesn't prominently post his affiliations.

The *National Post* is inclined to fall into the same trap. The paper carries opinion pieces by Milloy and by S. Fred Singer, again with no mention of the pair's corporate affiliations. Fred Singer is identified as the proprietor of the Science and Environmental Policy Project, an organization that sounds environmentally conscious. But readers might judge his input differently if the *Post* were more forthcoming about his corporate affiliations. The *National Post* can't fall back on a defense of ignorance, either. In an age when a Google search will turn up exhaustive details on Singer's background in milliseconds, readers should be able to expect that newspaper editors will check the bona fides of their contributors and share the details with their readers.

The *Calgary Herald*, another property in the Canwest Global Communications empire, is another offender. Calgary is Canada's oil capital: it hosts the head offices of Canada's major oil companies and is the economic (though not political) capital of Alberta, which with the tar sands boasts the world's largest known supply of oil outside the country of Saudi Arabia. Oil has made Alberta the wealthiest province in Canada, which cannot help but influence the policies of the province's leading newspaper. But that alone can't explain why the *Herald* would employ Friends of Science financier and University of Calgary professor Barry Cooper to write a weekly column without making any obvious effort to ensure that his biases don't overwhelm the facts.

Any newspaper is entirely within its rights to load up its opinion page with outspoken partisans, and Cooper is a good candidate for such a position. He is politically well-connected, and he is well-informed. But it is equally the newspaper's responsibility to ensure that the passionately expressed views of its most enthusiastic columnists rest roughly on a factual basis. Cooper makes outrageous accusations, saying that scientists are faking claims of climate change so they can fleece governments for additional research funding. This is one of the favorite claims of the denier community, and a clever one at that. They turn the accusation of self-interest—to which they themselves are vulnerable—on a community of international scholars who have chosen to pursue science rather than to enrich themselves in the service of industry. It seems that if he is going to continue bringing this up in the *Herald*, the newspaper might from time to time remind its readers that Cooper himself has benefited financially from oily investment in disinformation. In the circumstances, Cooper gets to climb onto his *Herald* soapbox, dismissing climate change as some kind of a complex hoax, without acknowledging the steps he took in the past to conceal from the public the oil-and-gas-industry funding for the so-called Friends of Science.

If it is wrong to open your pages to biased campaigners without admitting the nature of their bias, it is worse to actually commission work that appears crafted to confuse. Such was the case with the *National Post:* shortly after we at the DeSmog Blog began publicly compiling what we then called our "Denier Database"—a sort of rogues gallery—the *Post* launched a whole series of articles in 2007 called The Deniers. Week after week the periodic contributor Lawrence Solomon filed a column on a new scientist somewhere in the world whose work, according to Solomon's interpretation, somehow brought into question the main body of climate change science.

The problem with the *National Post* series was that Solomon's scientists were not as a rule denying anything. They were pursuing interesting research, some of which added to the complexity of climate change science, but in almost every instance, those scientists were also confirming that the world is warming and that human activity is to blame. One such scientist was Nigel Weiss, featured in a January 12, 2007, Solomon column titled "Will the Sun Cool Us?":

> "The science is settled" on climate change, say most scientists in the field. They believe that man-made emissions of greenhouse gases are heating the globe to dangerous levels and that, in the coming decades, steadily increasing temperatures will melt the polar ice caps and flood the world's low-lying coastal areas.

> Don't tell that to Nigel Weiss, Professor Emeritus at the Department of Applied Mathematics and Theoretical Physics at the University of Cambridge, past President of the Royal Astronomical Society, and a scientist as honored as they come. The science is anything but settled, he observes, except for one virtual certainty: The world is about to enter a cooling period.

The newspaper was hardly in the streets when Weiss started fielding phone calls and emails from family members and colleagues who knew that this was not an accurate interpretation of Weiss's work or of his personal views. And Weiss himself was stunned. He sat down immediately and penned a letter to the *National Post* demanding a correction and an apology. That letter, a copy of which he provided to the DeSmogBlog, read as follows:

> The article by Lawrence Solomon, which portrays me as a denier of global warming, is a slanderous fabrication. I have always maintained that the current episode of warming that we are experiencing is caused by anthropogenic greenhouse gases, and that global temperatures will rise much further unless steps are taken to halt the burning of fossil fuel. Compared to these effects, the influence of variations in solar magnetic activity is unimportant, however interesting it may be to astrophysicists like me. For further details see the Press Release on the University of Cambridge Web site.

If that wasn't clear enough, the press release mentioned above said, "Professor Nigel Weiss, an expert in solar magnetic fields, has rebutted claims that a fall in solar activity could somehow compensate for the man-made causes of global warming. Although solar activity has an effect on the climate, these changes are small compared to those associated with global warming. Any global cooling associated with a fall in solar activity would not significantly affect the global warming caused by greenhouse gases."

You might expect a responsible newspaper, having received this stunningly clear communication, to rush a correction into its next edition. Not the *National Post*. Instead, the editors passed Weiss's letter to Solomon, who wrote a querulous email in response. Weiss shared a copy of Solomon's email with Richard Littlemore at the DeSmogBlog on February 13, 2007:

> Dear Dr Weiss,
> I am writing in response to your letter to the editor of the National Post, in which you take issue with my characterization of you as a "denier." I can understand your objection to having a negative term assigned to you, although my use of this term is meant to be ironic—it is clear that I am not hostile to those I describe. I use the term "denier" to describe scientists who do not believe that the "science is settled."
>
> I do not understand, however, what other objections you may have to my column. Have I misrepresented your statements, for example, or attributed beliefs to you that you do not hold? If I have, I would very much like to understand how I have done so.
> Sincerely,
> *Larry Solomon*

Weiss, disappointed at this response, was even more surprised the next week, when Solomon wrote about him in another column criticizing people for using the term "denier" as an emotionally charged epithet. Weiss continued to complain—in fact threatened to sue—but the *Post* held out until April 17, 2007, three full months after the original article was published, before offering an apology and correction.

Solomon, however, continued to write columns under the banner The Deniers without ever enjoying much success in finding a scientist who in fact denies that humans are changing the world's climate in a dangerous way. That might seem like a surprising statement. It might even come across as hyperbole. Surely no one could go on, week after week, writing a column called The Deniers without including *deniers* in the column.

But I stand my ground and call as a witness Lawrence Solomon. Early in 2008 Solomon published a book version of this collection, called *The Deniers: The World-Renowned Scientists Who Stood Up against Global Warming Hysteria, Political Persecution, and Fraud (and Those Who Are Too Fearful to Do So).* It's a mouthful of a title, suggesting a combative and proud band of free-thinking scientists who have somehow endangered themselves and their careers by speaking frankly about their work—and by *denying* climate change. But on page forty-five Solomon says, "I also noticed something striking about my growing cast of deniers. None of them were deniers."

Solomon admits that none of his subjects were deniers. Not a single one. Even though Solomon quotes industry spokespeople such as Fred Singer and Bob Carter along with the more impressive characters within his book, Solomon acknowledged that he couldn't find a single serious scientist to actually deny the global consensus. I was surprised to find that admission within the pages of his book, even though Solomon tried to layer on confusion by saying that his subjects were "Affirmers in general. Deniers in particular." Essentially Solomon said that all of his protagonists deny *something,* it's just that none of them deny that humans are changing the state of the Earth's climate in a dangerous way.

I was yet more surprised that Solomon could go on for almost two hundred more pages, writing as though he had never made that fatal admission. But it's not one of the points that Solomon brings up on the lecture circuit or in the radio and television interviews I have seen since. He just sits proudly as they introduce him as the author of a newspaper series and a book about eminent climate change deniers. And I doubt that he was bringing it up in New York, where he too was appearing as one of the guest speakers at the "2nd International Conference on Climate Change."

I CAN COME up with no reasonable explanation for why the *National Post* and the *Calgary Herald* would allow this kind of material to continue to appear in their pages without taking some care to assure that readers were being fairly informed.

That isn't to say that the standard of journalism in Canada or elsewhere is uniformly low. Many of the stories within this book have emerged because of the diligence—and sometimes the courage—of great reporters. I am extremely proud of the DeSmogBlog team, especially Richard Littlemore and Kevin Grandia. And there are legions of reporters who, regardless of the standard that prevails in the media generally and at certain outlets specifically, do exemplary work. You can hardly find a better example than George Monbiot, *Guardian* columnist and author of the excellent 2007 climate change solutions book *Heat*. Monbiot broke one of the all-time-great climate disinformation stories—first in the *Guardian* and later in his book.

Monbiot, who had long since accepted the scientific consensus that he should be worrying about climate change, had stumbled in 2005 upon one of those striking pieces of information that would cause any reasonable person to have second thoughts. David Bellamy, at one time something of a giant in the U.K. environmental movement, had written a highly skeptical piece in the reputable *New Scientist* magazine in which he included this piece of information: "555 of all the 625 glaciers under observation by the World Glacier Monitoring Service in Zurich have been *growing* since 1980."

Giving a thought to Bellamy's reputation, Monbiot wrote in a May 10, 2005, *Guardian* column headlined "Junk Science," "[Bellamy] is a scientist, formerly a senior lecturer at the University of Durham. He knows, in other words, that you cannot credibly cite data unless it is well-sourced. Could it be that one of the main lines of evidence of the impacts of global warming—the retreat of the world's glaciers—was wrong?" Bugged by the numbers, Monbiot looked around for corroboration and, when he couldn't find it, picked up the phone and called the World Glacier Monitoring Service, where he encountered a helpful, well-informed individual who was familiar with the Bellamy story. Monbiot explained in the *Guardian* column how the monitoring service handled his query: "I don't think the response would have been published in *Nature,* but it had the scientific virtue of clarity. 'This is complete bullshit.'"

Now Monbiot was completely intrigued, and he set off on a sleuthing extravaganza, following every possible lead to try to find the source of this erroneous story. He found a variation of the figures in print: the libertarian Lyndon Larouche's publication *21st Century Science and Technology* had carried an earlier story stating that "55% of all the 625 glaciers under observation" by the World Glacier Monitoring Service were declining—suggesting that Bellamy had a source but rendered the reference badly. Monbiot also found similar references and variations littered around the Internet echo chamber—at JunkScience.com, where it can still be found; at the National Centre for Public Policy; at the industry-funded denier site GlobalWarming.org; and (this was the oldest link) at Dr. S. Fred Singer's Science and Environmental Policy Project Web site, sepp.org.

Monbiot stayed on the case. Singer had cited an old *Science* article from 1989, so Monbiot checked copies of the journal from the whole year. Nothing. Next, an associate phoned Singer directly and asked him the origin of the piece. Singer lashed out, saying that Monbiot "has been smoking something or other," and signed off, but in a subsequent conversation he broke down and admitted that, yes, the information had originated on his Web site. It had been posted there, Singer said, "by former associate Candace Crandall."

You might remember Candace Crandall's name from Chapter 4. She was one of the authors of the American Petroleum Institute's "Global Climate Science Communication Action Plan." Fred Singer should have had an easier time remembering her name: he's married to her.

While much of the criticism in this chapter falls at the feet of mainstream media, this story illustrates one of the problems of the Internet age. It is often difficult to assess the quality of information in a Web site story, especially if the proprietors of the Web site have taken the time or spent the money to create a professional-looking design. But it's also a great reminder, whether you are dealing with traditional outlets such as the *Guardian* or Internet sites ranging from JunkScience.org to the DeSmogBlog, you can never give up the ultimate responsibility for fact-checking. If someone tells you to be skeptical, be skeptical of them. For that matter, be skeptical of me. Search out credible corroboration for everything you read or hear, looking always to the credentials and the economic interests of those who are offering easy answers.

IT'S A DIFFICULT world out there for journalists. It must have taken Monbiot weeks to finally track down Singer's glacier story, and few reporters have that kind of time, the freedom to phone people all over the world, and the necessary attention span. In a business that calls upon you to fill a daily void, most reporters wind up running ragged just to meet the responsibilities of a given day. The situation has also gotten worse as major media have consolidated. Although there are more individual outlets in North America today, the total number of reporters has been declining for more than twenty years. People are working harder, specialists are stretched thinner, and media empires like Canwest Global are always looking for more ways to repurpose the same information in more products and more markets.

It's little wonder that reporters fall back on quoting one source on one side of an issue and one on the other. It's little wonder that they tend to stand aloof as long as they can, trying not to get suckered into committing themselves on a controversial story until there is absolute agreement from all sides. Doing so on an issue like climate change is especially risky when media managers have a closer relationship with auto advertisers than they do with scientists at the local university.

Those scientists also share some of the blame for the continuing public confusion, according to Naomi Oreskes. Scientists are a media-shy group, and for good reason. The scientific fraternity is generally hostile to anyone who courts media attention—especially someone who leaks interesting research results to mainstream media before that research has been vetted in a formal peer-review process at a respectable scientific journal. And while PR-trained professionals such as Tim Ball and Fred Singer snap off one quotable line after another, scientists—sincerely skeptical and devoutly specific—will often couch their comments in carefully conditional language that leaves people wondering what exactly has been said. Consider, for example, the most recent report of the Intergovernmental Panel on Climate Change, in which the best scientists in the world worked themselves up to saying that anthropogenic (which is to say, human-induced) climate change is "very likely." Tim Ball, in the meantime, is writing sentences like this one, from the Web site CanadaFreePress.com: "Believe it or not, Global Warming is not due to human contribution of Carbon Dioxide (CO_2). This in fact is the greatest deception in the history of science." Nothing conditional about that. It's clear and concise, and it arrives with a memorable punch. Think of yourself as a young reporter—an arts major who dropped physics in grade 11—trying to decide whose quote you should use in the lead of your story. It's no wonder, given the certainty with which it is packaged, that uncertainty continues to sell.

But Oreskes's criticism is slightly different. She says she and her colleagues are inclined to practice "supply-side science," counting on a "trickle-down effect" and relying on a "diffusion model" to get the word out. These are intentionally funny economist-talk descriptions, but they get to a serious point. Knowledge does not redistribute itself by osmosis. It doesn't flow automatically from the areas of greatest concentration, usually universities, to the neighborhoods that need it most. Knowledge has to be distributed with a sense of purpose and often a talent for engaging your audience. That's why public relations people have jobs: we specialize in communicating the kind of information that might get overlooked if our clients didn't take the initiative.

The good news about mainstream media generally is that they seem to be moving in a better direction. When Max Boykoff followed up in 2006 on the research that he and his brother had published in 2003, he wrote a paper called "Flogging a Dead Norm? Newspaper Coverage of Anthropogenic Climate Change in the United States and United Kingdom from 2003 to 2006." Boykoff found that most media outlets had abandoned the balance model in covering climate change. No longer were they regularly soliciting (or allowing) a quote from "the other side" every time they ran a story about global warming.

But it hasn't stopped altogether. While working on this book, I took a break one morning and flicked on the BBC World Service. The British network offers a different perspective on international news than you get watching Canadian or American television, and the BBC was one of the first networks to declare as an organization that they accepted the science of climate change and no longer felt compelled to accommodate an argument in every story. But here, in February 2009, was a presenter introducing a climate change story with this self-censoring aside: "Although some scientists say that humans may not be causing global warming."

I wanted to scream at the television: That's not true! If Benny Peiser can't find a single peer-reviewed article in any reputable science journal any time in the last fifteen years, if Lawrence Solomon can't find even one well-qualified "denier" who in point of fact *denies* the human contribution to potentially dangerous climate change, well, this alleged scientific controversy can only be dismissed for what it is—a carefully constructed ruse to keep people from supporting the kinds of actions that will compromise the profit potential of ExxonMobil, the Western Fuels Association, and the American automakers, whose fortunes were shattered after they bet their futures on the continued gullibility of the SUV-buying public.

It's not true, and it's past time for people in the media to check their facts and start sharing them, ethically and responsibly, with a public that is hungry for the truth.

[*thirteen*]

MONEY TALKS

Calculating the heavy weight of political capital

There is a quaint assumption in North America that corporations invest in democratic politics out of a sense of public spiritedness. It would not fit with our egalitarian illusions to believe that people with a lot of money could just buy politicians and then order up whatever policy served their interests, and such a belief would demean politicians to the point that their current reputation (above car thieves; below car salesmen) might decline even further.

Traditionally, major corporations and industry associations have tried to encourage this mythology by keeping their contributions relatively even among the major parties. In the United States, this old habit has given belligerent independents like Ralph Nader an important talking point: that the Democrats and Republicans are (or have been) equally beholden to their corporate donors. But it also offered a vague sense of reassurance that political contributions were somehow nonpartisan. The situation is a little different in Canada, where you have a socialist/labor contingent in the form of the New Democratic Party. But generally, here too corporations honor the tradition of distributing their political largesse roughly equally between the two major parties (the Liberals and the Conservatives).

Within the energy industry the tide started to shift in the early 1990s, most obviously in the United States and most definitely away from the Al Gore Democrats. If you look at the fabulous records kept at OpenSecrets.org by the Center for Responsive Politics, you notice that oil-and-gas-industry contributions to U.S. politicians, which were divided by less than 60/40 between Republicans and Democrats in 1990, have swung toward the Republicans, reaching 75/25 during the later Clinton years and rising to 80/20 under the Bush administration. President George W. Bush himself received more than US$1.5 million in direct contributions from the oil industry in the run-up to his first presidential election—at the time, the largest amount of money the industry had ever given to a presidential candidate.[1]

Other sectors that are heavily invested in fossil fuels followed suit. Contributions from coal-mining firms, also sitting around 60/40 for Republicans versus Democrats in 1990, favored Republicans 90/10 by the middle Bush years. This was a particularly fortuitous influx of support for Bush in the controversial 2000 election. By winning five electoral votes from the traditionally Democratic coal state of West Virginia, Bush advanced to within a Supreme Court decision of the president's office.

The same movement of political cash can be tracked in the spending of electrical utilities (almost all coal fired) and other fossil fuel–dependent industries such as the automotive sector. Industries that had always more or less split the difference were suddenly and dramatically favoring the Republicans.

We're talking serious money. In 2004 the Center for Public Integrity reported that the oil-and-gas industry had spent more than US$420 million on lobbying and political contributions in the preceding six years. ExxonMobil alone had spent US$60 million. The numbers were similarly large in the other energy sectors. According to Jeff Goodell, author of the 2006 book *Big Coal,* the coal-fired electrical giant Southern Company spent US$25 million on lobbying between 2001 and 2004 and an additional US$4.4 million in political contributions—a single, comparatively small utility company outspending much larger corporations such as Ford, Pfizer, and Monsanto.

George Bush might argue that it was a total coincidence that one of his administration's first decisions was to walk away from its responsibilities under the Kyoto Protocol, the agreement negotiated in 1997 under the UN Framework Convention on Climate Change that committed developed nations (and the biggest polluters in the world) to reducing their carbon dioxide emissions by set amounts before 2012. Other countries, most notably Canada, were failing to meet their targets, but among the major players only the United States walked away from the table altogether.

Bush also set about dismantling U.S. EPA pollution regulations. For example, Goodell reports in *Big Coal* that under Bush the EPA launched the 2002 Clear Skies Initiative, which, as originally written, would have cost the coal-fired electrical utility industry US$6.5 billion to implement but would have produced US$93 billion in health benefits. But Clear Skies backed away from specific pollution limits on individual plants, moving instead to a cap-and-trade system—so it would have functionally relaxed some regulations that had been passed, but not enacted, under the Clinton-era Clean Air Act. Clear Skies also completely ignored carbon dioxide as a problematic waste substance, even though coal-fired electrical utilities are the biggest source of greenhouse gas production in America—and America was then the biggest producer of those gases in the world.

Still, Clear Skies would have forced the electrical utility industry to reduce dangerous mercury emissions from forty-eight tons to twenty-six tons by 2010, and to fifteen tons by 2018 (bear in mind that *no* level of mercury is considered safe for humans). But over the next year of lobbying, Clear Skies faltered in Congress, and the Bush administration moved to dismantle the tougher Clinton standards through regulations instead. By the time the new standards passed, the EPA had unilaterally removed coal plants from the list of those facilities governed as sources of hazardous air pollutants. And the mercury limits had jumped up from twenty-six and fifteen tons to thirty-four and twenty-six tons in 2010 and 2018, respectively. Asked about the reason for these changes, Bush officials answered, "The EPA, in its expert judgment, concludes that utility [mercury] emissions do not pose hazards to public health." But as Goodell writes, the *Washington Post* reported a few weeks later that the "expert judgment" had come directly from industry: "At least a dozen passages in the EPA's proposal were lifted, sometimes verbatim, from memos prepared by West Associates, an industry organization representing western coal burners, and Latham and Watkins, a powerful Washington law firm that often represents corporations on environmental issues." There was, in other words, compelling evidence that the EPA was taking direction from precisely the people it was supposed to be regulating.

While gutting environmental restrictions and promoting oil drilling (or as Frank Luntz would prefer, "energy exploration") in the Arctic National Wildlife Refuge, the Bush minions were also rewriting reality in the White House. Rick Piltz, who was then a senior associate in the U.S. Climate Change Science Program, described in a recent interview with John Mitchell and Jana Christy for the blog shuffleboil.com the context of scientific discussion and the Orwellian twists that prevailed at the time:

> The science community tends to have a lot of integrity about "Here are the questions we still don't have answers to," and they're up front about where they're uncertain . . . But if you come into that with a predatory relationship to uncertainty, and you say, "Well, they have all these questions they haven't resolved yet, there's all this uncertainty, and we have a scientist here who thinks something different from that scientist there, so obviously there's still some big debate and, of course, we can't do something about [it] in terms of policy until all these scientific issues are resolved." That's what I saw the Bush administration White House people doing with the science program reports that I was working on. They were adding and deleting in such a way as to systematically—paragraph after paragraph, page after page—introduce the idea that there was some sort of fundamental scientific uncertainty that still needed to be debated and they would seize on any stray factoid or study or think tank, or whatever, in order to do that. And it was totally political. It didn't have anything to do with science and the way scientists think and the way the scientific literature develops. It was totally predatory.

The key word here is "systematically." Piltz is accusing the Bush administration of corrupting the public record intentionally and continually—and the accusation stood unchallenged. That's not to say that the Republicans passed up the opportunity to attack Piltz when he resigned in 2005. But several months later, on June 8, 2005, Andy Revkin at the *New York Times* reported in a story titled "Bush Aide Softened Greenhouse Gas Links to Global Warming" that he had received copies of documents that had been subjected to just that kind of editing.

The perpetrator was Philip A. Cooney, the chief of staff for the White House Council on Environmental Quality—a position that he had received despite having no background in science. He was an economist and lawyer, and his qualification—at least for the purposes of the Bush administration—was that he had been the "climate team leader" and a lobbyist at the API. In a series of reports that passed over his desk in 2002 and 2003—science reports that had been created and vetted by government scientists, their supervisors, and, in some cases, senior Bush administration officials—Cooney had made specific editorial changes, deleting details that would raise concern about climate change and adding adjectives to play up uncertainty.

Andy Revkin reported examples in the *New York Times* such as

> the insertion of the phrase "significant and fundamental" before the word "uncertainties." In an October 2002 draft of a summary of government climate research, *Our Changing Planet*, Cooney added the word "extremely" to this sentence: "The attribution of the causes of biological and ecological changes to climate change or variability is *extremely* difficult." In a section on the need for research into how warming might change water availability and flooding, he crossed out a paragraph describing the projected reduction of mountain glaciers and snowpack. His note in the margins explained that this was "straying from research strategy into speculative findings/musings."

Responding to Revkin, White House officials defended Cooney's edits, saying that they were part of the normal review that takes place on all documents related to global environmental change. But the most damning argument that Revkin reported from the administration's attempts to defend Cooney came from Myron Ebell, the director of global warming and international environmental policy at the CEI, who said that Cooney's editing was necessary for consistency in meshing programs with policy. Clearly, neither Cooney nor Ebell were interested in meshing programs or policy with a reasonable and accurate version of the science.

Ebell's interjection served as a reminder that CEI had also featured in an earlier scandal, in which Greenpeace had turned up a 2002 memo from Ebell to Cooney, thanking Cooney for inviting CEI to help the Bush administration undermine the efforts of the EPA. By the time the memo came to public attention, CEI had also sued the administration, invoking obscure provisions of the Federal Data Quality Act in an effort to block the public release of the *National Assessment of the Potential Consequences of Climate Variability and Change.* In an interview with the DeSmogBlog in 2006, Piltz said he was convinced that if the *National Assessment* had been released and had received the attention that it deserved, the administration would have no longer been able to deny the dangers of climate change or continue putting off action to address it.

In the land of happy endings, immediately after Revkin reported the illicit edits, Cooney submitted his resignation, purportedly to spend more time with his family, and within a week took up full-time employment with ExxonMobil.

The White House continued pursuing its editorial policy the following year, slashing by more than 50 percent the testimony that Centers for Disease Control director Julie Gerberding was permitted to give to the Senate Committee on Environment and Public Works. In response to White House demands for an advance copy, Gerberding had submitted a draft of her testimony to the White House before appearing at the committee hearing in October 2007. As reported at the time by MSNBC, "White House press secretary Dana Perino said the prepared testimony went through an interagency review process and the Office of Science and Technology Policy did not believe that the science in the testimony matched the science that was in a report by the International Panel on Climate Change . . . [Perino said of the report] 'It was not watered down in terms of its science. It wasn't watered down in terms of the concerns that climate change raises for public health.'"

But when Kevin Grandia at the DeSmogBlog obtained a full text of the proposed testimony, it became clear that the White House had slashed the testimony in half, removing almost all references to negative health effects and allowing Gerberding only to describe to the committee the measures that the CDC had made to prepare itself for a climate-changed America.

Probably the highest-profile accusation of Bush administration censorship concerned the scientist James Hansen, for almost thirty years the director of NASA's Goddard Institute for Space Studies. In sharp contrast to my earlier description of scientists as media-shy and constantly making understatements, Hansen has been one of the clearest and most outspoken climate commentators in the world for more than two decades. Hansen first caught government attention in 1988, when he appeared before the Senate Energy and National Resources Committee saying, "It is time to stop waffling so much and say that the evidence is pretty strong that the greenhouse effect is here."[2]

Hansen's position and the quality of his contribution to climate science have made it difficult to criticize him, though the denier community has tried time and again. For example, California Republican Darrell Issa accused Hansen in 2007 of being a partisan supporter of the Democrats and, especially, of John Kerry in his challenge for the presidency against George Bush in 2004. Issa launched the criticism on the basis of a US$250,000 science prize that Hansen had won from the Heinz Foundation. The Heinz Awards are given annually to honor individuals working in areas important to the late senator John Heinz, a Pennsylvania Republican. The prize is awarded by Heinz's widow, Teresa Heinz, who is now married to Kerry.

Hansen himself had complained in 2006 that George Deutsch, whom President Bush had appointed as a NASA press aide, was trying to limit or control what Hansen and other scientists were saying to the media. Hansen told *60 Minutes,* "In my more than three decades in the government, I've never witnessed such restrictions on the ability of scientists to communicate with the public." Hansen said that NASA scientists were told that their Congressional reports and public presentations must be reviewed and approved beforehand, and that they could only participate in media interviews if a NASA public affairs staffer sat in. Deutsch later denied the efforts at censorship but was forced to resign over a scandal that broke when it was discovered, first, that he had ordered a NASA Web site designer to add the word "theory" beside every reference to the big bang, and second, that he had lied on his resume, claiming a university degree that he had never obtained.

And despite Deutsch's insistence that Hansen had overstated the efforts to muzzle him, in 2008 NASA released a forty-eight-page report arising out of an internal investigation into the charges. The report concluded, "During the fall of 2004 through early 2006, the NASA Headquarters Office of Public Affairs managed the topic of climate change in a manner that reduced, marginalized or mischaracterized climate change science made available to the general public." Despite the clarity of this finding—a flat statement that NASA Public Affairs was corrupting the flow of scientific information to the public, who had paid for the research—the story blew over quickly, with little coverage in most major media and a shrug of recognition from those closest to the issue.

Not to be outdone in the censorship department, the Canadian government took a clumsy stumble into the game in 2008 and got caught issuing an embarrassing "Media Relations Protocol," a PowerPoint presentation now available on the DeSmogBlog, that was transparently designed to frustrate or prevent media relations. The new rules, distributed to all Environment Canada scientists, said that no one was allowed to respond to any reporter's query before first consulting with their supervisor. If it was then judged appropriate to respond to the question, it would be referred to the Media Relations office in Ottawa, where political staff could "respond with approved lines." This was designed to overcome two supposed problems. The protocol held that there had been "limited coordination of messages across the country" and that Environment Canada "interviews sometimes result in surprises to [the] minister and senior management."

Despite the similarities in the styles of climate change positioning that have prevailed in Canada and the United States, especially during the later years of the Bush administration, the direct dollar-driven corporate influence over politicians isn't as easy to establish in Canada as it is south of the border. The political system is different, and Canada's limits on corporate donations are tighter, especially recently, but there has never been any doubt as to the domination of corporate advantage over environmental responsibility in the making of Canadian climate policy. At least, that's a reasonable conclusion when you realize that at the federal level especially, there was no Canadian policy in place as of early 2009 to actually reduce national greenhouse gas emissions. The Conservative Party administration of Prime Minister Stephen Harper is still talking about slowing the rate of increase.

Yet Canada once had a pretty good reputation on environmental issues. The treaty that ultimately turned around the destruction of the ozone hole was signed in Montreal in 1987. The first international conference on climate change was held in Toronto in 1988, and when the world gathered in Rio de Janeiro in 1992 for the Earth Summit, Canadian prime minister Brian Mulroney was the first leader to sign the resulting UN Framework Convention on Climate Change.

But Canada was deeply in debt in the early '90s and subject to a gathering antigovernment sentiment. Although Mulroney's nominally Conservative government had dug the debt hole deeper, Liberal Party prime minister Jean Chrétien, who was elected in 1993, immediately set about cutting government spending and pursuing his election promise of providing "jobs, jobs, jobs" for Canadians, with no evident regard for a deteriorating environment. By 1997, when the UN Framework Convention on Climate Change partners were negotiating a new climate protocol in Kyoto, Canada had joined the corporate resistance, but grudgingly accepted the challenge to cut national greenhouse gas emissions to 6 percent below 1990 levels by 2012.

Unfortunately, there was never a possibility that Canada would meet that commitment. In 1998 Prime Minister Chrétien set up a cumbersome and cynical "implementation process." Consisting of sixteen "tables" incorporating every subject area and interest group in the Canadian political panoply, the process was doomed from the start. The featured speaker at the kickoff conference was not the prime minister, or even his minister of the environment, but rather the deputy minister, dispatched to try to assure the assembled lobbyists and environmentalists that the government was taking the process seriously.

In the pitched negotiations that followed, the corporate representatives from organizations like the Canadian Association of Petroleum Producers and the Cement Association of Canada brought a three-part argument. First, they said, climate change might not be happening, and if it is, humans might not be to blame, so why bother wrecking the Canadian economy in an attempt to fix it? Second, if climate change is happening, and humans are to blame, the Kyoto Protocol is a completely inadequate response—so why bother wrecking the Canadian economy in an attempt to implement it? And third, as Canada's economy is inextricably linked to the economy of the United States, and as the United States, even with Bill Clinton and Al Gore in the White House, clearly has no intention of meeting its commitments, why bother wrecking the Canadian economy when nothing we do can balance the damage being done south of the border?

We have already been talking about the first argument, and the third argument is still very much in vogue—except that now North American foot-draggers are more likely to point the finger at India and China when recommending that we all throw our hands up in despair and resume mining tar sands and coal as fast as we can. But in 1988 the second argument was the most infuriating, for two reasons: First, it was true. If the world community hopes to address the threats of climate change, then the provisions contained in the Kyoto agreement are woefully insufficient to the task. Second, by making this point, the corporate lobbyists demonstrated that they had read enough of the science to understand the size and scope of the problem. Yet they still recommended inaction as the response.

And they prevailed. Two years later the Liberal government produced its *Action Plan 2000 on Climate Change*, promising to spend just C$1.1 billion over five years and to implement changes that would reduce Canada's greenhouse gas emissions by seventy-two megatons a year in the Kyoto reporting period between 2008 and 2012. But the proposed reductions were always loosely calculated and seldom deducted directly from a current total. Far from reducing greenhouse gas emissions, Canada had been increasing its output by an average of more than sixteen megatons a year for the previous eight years. Instead of being 6 percent below 1990 levels, the country was closer to 20 percent above—which is to say that Canada was at nearly 30 percent over its Kyoto target and still moving in the wrong direction.[3]

Notwithstanding his record of inaction and the virtual impossibility of meeting Canada's commitment, Prime Minister Chrétien's government ratified the Kyoto Protocol in 2002 and implemented a handful of increasingly promising policies over the next three years. But the political winds had shifted from the eastern-based Liberal Party to the western-based Conservatives—a re-formed right-of-centre coalition that was not nearly as green as the Mulroney Progressive Conservatives of the late 1980s.

When Conservative leader Stephen Harper campaigned for the prime minister's job in early 2006, he did so on an interesting two-track campaign. He promised first of all to be a law-and-order conservative who would crack down hard on people who broke the law. And he promised to abrogate Canada's international legal commitment to the Kyoto Protocol. If any reporters noticed the contradiction, there is no record that they asked him about it. Once elected, the prime minister promptly handed the job of environment minister, which included defending the government's climate change position, to the Alberta member of parliament Rona Ambrose. Ambrose came across as a Canadian version of the benighted Sarah Palin—attractive, initially popular, and totally out of her depth, which turned out to have been a policy decision. Government scientists in Environment Canada reported to their private-sector colleagues that Ambrose declined to be briefed on the science of climate change.

The Conservative hostility to Kyoto surprised no one. Prime Minister Harper was himself elected as a member of parliament from the oil capital, Calgary (his actual riding is Calgary Southwest), and his party's base is preponderantly in resource-rich western Canada, and decidedly in oil-rich Alberta.

In 2006 Alberta was awash in cash. Between 1990 and 2006 fossil fuel industry revenues had climbed 61 percent.[4] But that windfall had come with a complicating factor: industry-source greenhouse gas emissions had increased over the same period by 53 percent, accounting for almost half of the total increase in emissions recorded over the same period. Most of the rest came from transportation and from coal-fired electricity generation.

Harper's position on climate change, that of a loyal Albertan, had been on the record—and perfectly unclear—for years. In a story published December 21, 2006 ("PM Denies Climate-Change Shift"), the *Toronto Star*'s Ottawa bureau chief, Susan Delacourt, chronicled the evolution of the prime minister's thinking. In September 2002, for example, he passed off the issue as a controversy of little interest to Canadians: "It's a scientific hypothesis, a controversial one and one that I think there is some preliminary evidence for . . . This may be a lot of fun for a few scientific and environmental elites in Ottawa, but ordinary Canadians from coast to coast will not put up with what this [the Kyoto accord] will do to their economy and lifestyle, when the benefits are negligible." In 2004, Delacourt writes, the prime minister updated that position to say, "The science is still evolving." And by 2006 he was still referring to "so-called greenhouse gases." If you give him the benefit of the doubt, Prime Minister Harper seemed, even as he took over the reins of power, to be like those well-educated Republicans from Chapter 12, so steeped in uncertainty that he couldn't bring himself even to believe in the existence of the greenhouse gases that Joseph Fourier had discovered in the early part of the 19th century.

In addition to announcing that he had no intention of trying to meet Canada's Kyoto targets, the Canadian prime minister also set about dismantling all the climate change policies that the previous Liberal government had implemented to date. He shut down the government's climate change Web site and removed all references to global warming, and especially to Kyoto, from federal communications, except to say that henceforth he would be resisting international pressure and pursuing a "made-in-Canada solution."

Here Harper begins to use language that was actually made in America. The Republican spin doctor Frank Luntz was in Kingston, Ontario, in May 2006, speaking to the Conservative-linked Civitas Society and making time on the side for a personal meeting with Prime Minister Harper. (The prime minister confirmed in the House of Commons a couple of days later that he and Luntz had been acquainted "for some years.")

In the weeks that followed, people started listening more closely to the Conservatives and looking for likely connections to the strategy document, as discussed in Chapter 6, that Luntz had written for the U.S. Republican Party ("The Environment: A Cleaner, Safer, Healthier America"). As Ross Gelbspan recorded on the DeSmogBlog on May 31, 2006, the *Kitchener Waterloo Record* reported the results in a story headlined "Tory Kyoto Strategy Mirrors U.S. Plan":

> In his 2003 memo, [Luntz] told Republicans not to use economic arguments against environmental regulations, because environmental arguments would always win out with average Americans concerned about their health. Luntz also told his U.S. clients to stress common sense and accountability. "First, assure your audience that you are committed to 'preserving and protecting' the environment but that 'it can be done more wisely and effectively.' Absolutely do not raise economic arguments first."
>
> Since the Conservatives took office, they have consistently stressed their commitment to clean air and water, and tried to avoid discussion of cutting back environmental programs—although many have been eliminated. "My mandate is to have accountability on the environment and show real results and action on the environment for Canadians," [Environment Minister] Ambrose told the Commons last week.
>
> Luntz advises that technology and innovation are the keys to curbing climate change, a theme the Conservatives

> have repeatedly echoed. "We will be investing in Canadian technology and in Canadians," Ambrose told MPs.
>
> Despite his general aversion to economic arguments, Luntz... advises putting the cost of regulation in human terms, emphasizing how specific activities will cost more, from "pumping gas to turning on the light." Ambrose has claimed that "we would have to pull every truck and car off the street, shut down every train and ground every plane to reach the Kyoto target. Or we could shut off all the lights in Canada tomorrow."

In this first year that the Harper Conservatives were in power, Canada was also the official chair of the UN Framework Convention on Climate Change, which gave the country two special chances to drag down the process. First, Environment Minister Rona Ambrose was anything but a champion for action. She dismissed Canada's own commitments, blew off Canada's reporting deadlines, and on one occasion at least, declined even to attend a meeting, assuming her position as "chair" over the telephone.

Canada also increased international inertia on behalf of the Bush administration. Having refused to sign the Kyoto Protocol, the United States was effectively sidelined from the process, forced to sit outside of the most critical meetings awaiting word of how the parties to the accord were planning to proceed. Given that the United States was the world's number-one producer of greenhouse gases, there was only so much that could be decided by the remainder of the world's powers, but the United States still feared that its interests could be marginalized by a concerted international effort to discourage emissions.

That was no threat with Canada in the room. Having backed away from its own Kyoto commitments, Canada also chose to join the Asia-Pacific Partnership on Clean Development and Climate, a sort of anti-Kyoto coalition that included the world's biggest polluters (China, India, Japan, Korea, and the United States) and the second-tier countries that sell them oil and coal (Australia and Canada). Even Republican Senator (and later presidential candidate) John McCain dismissed the partnership as "nothing more than a nice little public relations ploy." McCain told Grist writer Amanda Griscom Little on August 4, 2005 ("New Asia-Pacific Climate Pact Is Long on PR, Short on Substance"), that the partnership had "almost no meaning. They aren't even committing money to the effort, much less enacting rules to reduce greenhouse gas emissions." The group's apparent determination to create an alternative organization that could be used to undermine the UN Framework Convention on Climate Change, combined with the Canadian decision to join the Big Coal coalition during a year when Canada was nominal chair of the UN Framework process, dealt the UN body a telling blow.

At subsequent UN Framework conferences, especially in Bali in 2007, Canada's obstructionist position became so obvious that people started to believe the Bush and Harper administrations were working together—that Canada was trying to prevent any progress that might demonstrate how badly the United States was out of step. But the theory broke down at the Kyoto Protocol update in Poznan, Poland, in December 2008. By then Barack Obama was already president-elect, though George Bush would retain the actual presidency until January 20, 2009. So the Bush negotiators were still in the room, but with no real mandate: everyone expected that the Obama administration would take a more aggressive tack in approaching climate change.

With the United States removed as a contrarian force, some people expected that Canada would shift to a more productive position as well. But if anything, Canada stepped up its obstructionism, urging other countries to back away from greenhouse gas reduction commitments they had made in Bali the year before. For its efforts Canada was granted the "Colossal Fossil" award. The environmental Climate Action Network chose a "Fossil of the Day" for each day of the two-week conference, and the country with the most nominations was judged to be the Colossal Fossil when the meeting wound down. Canada really earned that international embarrassment.

While dragging down efforts to build an effective greenhouse gas reduction policy on the world stage, the Harper Conservatives continued to emulate U.S. policy at home. Where in 2003 the Bush administration had proposed a Clear Skies Act that ignored greenhouse gases almost entirely, the Harper Tories followed with a Clean Air Act in 2006, which focused on smog and particulate pollutants and promised (still voluntary) emission targets by 2020.

With U.S. president George Bush advocating "energy intensity targets" as a way to address climate change, the same policy started appearing in Canadian climate documents only a short while later, such as in Environment Canada's 2008 regulatory framework for industrial greenhouse gas emissions. An energy intensity target is something you might expect to get from Frank Luntz: it's very specific. The definitions are clear and concise. But when you implement it succesfully, you get a public relations boost without any corresponding reduction in greenhouse gas emissions. Consider, for illustration, the following definition from the World Resources Institute: "Greenhouse gas intensity targets are policies that specify emissions reductions relative to productivity or economic output, for instance, tons CO_2/million dollars GDP. By contrast, *absolute emissions targets* specify reductions measured in metric tons, relative only to a historical baseline." That means that you can reduce energy intensity by a lot (the Canadian tar sands giant Suncor cut its energy intensity by 51 percent between 1990 and 2006) while at the same time continuing to make the problem worse (despite the "intensity" cut, Suncor increased its absolute emissions by 131 percent during the same period).

Thus, intensity targets are for people who don't want to deal with the problem. Consider this May 7, 2001, statement from Bush White House spokesperson Ari Fleischer in response to a question about whether the president would urge Americans to change their world-leading energy-consumption habits: "That's a big 'no.' The president believes that it's an American way of life, that it should be the goal of policy-makers to protect the American way of life. The American way of life is a blessed one . . . The president considers Americans' heavy use of energy a reflection of the strength of our economy, of the way of life that the American people have come to enjoy." True to his word, until oil prices spiked in the summer of 2008, the Bush administration held its position, touting energy intensity cuts while supporting the expansion of the coal-fired power industry and the aggressive extension of oil drilling into parks and oceans.

Here's how things played out in Canada during the same period: the provincial administration in Alberta, home to the largest section of Canada's huge tar sands deposit, announced a climate change strategy in 2008 that would call for no greenhouse gas emission reductions whatsoever before 2020. In a document titled *Responsibility/Leadership/Action,* Alberta also proposed to pursue energy intensity targets in the short term (2010), to "stabilize" emissions by 2020 and to "reduce" emissions by 2050 by 14 percent from 2005 levels. Put another way, Alberta was planning to give industry free rein until 2020, after which it would introduce regulations so gently that by 2050, the province still would not comply with the target that Canada promised in Kyoto to meet by 2012. Returning once again to the dark definition of Orwellian, it's hard to imagine how that could seriously be described as responsibility, leadership, or even action.

Nationally the Harper government came out in 2008 with its regulatory framework for greenhouse gas emissions. This new plan, dubbed *Turning the Corner,* set out a strategy that would keep the country straight on course. It concentrated on energy intensity targets even while allowing unrestricted tar sands development at least through 2012. New coal-fired power plants would get a free pass for at least a decade. They would have to be "carbon-capture ready" by 2018, but there was no hard deadline for when they might actually have to start capturing carbon. Nevertheless, the government still imagined that it could reduce total national greenhouse gas emissions by 20 percent below 2006 levels by 2020. Keeping in mind Canada's record of increasing greenhouse gas emissions between 1990 and 2006, the Harper government had set out a plan that would violate its Kyoto commitments for at least another dozen years.

This again is not leadership. Rather, in Canada and in the United States, it looks just like a bunch of politicians overlooking a serious scientific issue and ignoring the interests of the public in favor of serving the interests of a wealthy and well-connected industry group. It's also clear that the politicians on both sides of the border know that what they are doing is wrong. How else can you explain the use of a policy like energy intensity targets?

If the governments truly believed that addressing climate change was unnecessary—if these politicians were confident that they were doing the right thing by pursuing unrestrained, old-style industrial development at increasing risk to the environment—there would be no reason to rustle up phony carbon cuts to make it sound like they were taking the climate issue seriously. If the provincial and federal governments in Canada were not wrestling with a guilty conscience, why would they come up with promises to reduce greehouse gases by 14 percent by 2050 or by 20 percent by 2020—all the while changing the base year so you can't compare one proposal with the next?

EARLY IN MARCH 2009 DeSmogBlog manager Kevin Grandia started poking through the public records at OpenSecrets .org, looking at how much money the oil-and-gas industry had been spending on political lobbyists in the United States. Grandia reported his findings in a March 17 blog post, the headline for which did a good job of telling the story: "Oil and Gas Lobbying on Capitol Hill Up a Whopping 57% in 2008." The stunning part is that the oil-and-gas-industry lobbying budget was already US$82 million in 2007, which means the companies threw in an additional US$46.6 million, to bring the 2008 total to US$128.6 million. In his report, Grandia also quoted from a Center for Public Integrity story called "Climate Change Lobby Explosion": "Senate lobbying disclosure forms show that more than 770 companies and interest groups hired an estimated 2,340 lobbyists to influence federal policy on climate change in the past year, as the issue gathered momentum and came to a vote on Capitol Hill. That's an increase of more than 300 percent in the number of lobbyists on climate change in just five years, and means that Washington can now boast more than four climate lobbyists for every member of Congress."

Four lobbyists for every member of Congress. Does that seem excessive to you? That group will include a significant minority of lobbyists from environmental groups that are trying to get politicians to take this issue seriously, but the oil-and-gas industry's US$18.6 million is certainly paying for the vast majority of these influencers. And you have to assume that the industry believes it is getting value for money. Why, otherwise, would it increase its spending by 64 percent in a single year?

We know, then, who's looking after the interests of Big Oil. There are big, burly lifeguards aplenty in that camp. And once again, it's a free country. The oil companies have every right to do what they can to protect their profits. Or maybe not. Maybe it's wrong to try to deny the science of climate change even after your own scientists have assured you that the case is undeniable. Maybe it's wrong to bury that report and then organize and implement a strategy designed to hold people "balanced" in confusion. And maybe the politicians who have asked for our votes should be closing the doors on a few of those 2,340 lobbyists and taking the time to read some actual science—and then to act in the public interest. Maybe it's time we all stood up and demanded a little equal time before these people direct us all off a cliff onto the ever-more-undeniable rocks waiting below.

[*fourteen*]

WHITEWASHING COAL

In coal country, cleanliness is relative, but profit is absolute

From the destruction of green Welsh valleys to the clogging of lungs in old London, coal has long been seen as an anti-environmental culprit, a dated and clumsy fuel source that was dangerous to mine, difficult to move, and dirty to burn. In the course of the 20th century, one industry after another discovered the superiority of fuels like diesel and natural gas, and coal seemed to slip into the background as an energy source whose time had passed.

Or perhaps not. For people living in the disappearing Appalachian Mountains, coal's currency is an unavoidable reality. As Jeff Goodell reports in *Big Coal*, mining by mountaintop removal has destroyed more than seven hundred miles of streams and "turned about 400,000 acres of some of the world's most biologically rich temperate rainforest into a flat, barren wasteland." After a century in which coal mining accidents killed 100,000 people and black lung claimed the lives of 200,000 more, the death toll continues. In January 2006, for example, seventeen men died in Appalachian coal mines.

Coal's resurgence—or maybe it's just persistence—is based on three things. First, some of the biggest and politically most influential players in the United States, from coal mining firms to railroads and electrical utilities, are heavily invested in coal and don't want to give it up. Second, at a time when the world's foremost experts are warning about "peak oil," the coming collapse in supplies of both conventional oil and natural gas, the world supply of coal is to some a reassuring backstop. According to *Big Coal*, the United States alone has known coal reserves amounting to more than 270 billion tons, enough to meet current U.S. consumption for the next 250 years. Third, America has led the world with its de facto declaration that given a choice between short-term profits and the long-term viability of the planet, profits rule. So even though coal-fired electrical plants already produce 40 percent of all U.S. greenhouse gases, and even though that makes the United States the largest national source of greenhouse gases in the world, the current strategy is to continue mining, moving, and burning coal at a rate redoubled from the dangerous past.

The implications are immediately obvious. When burning coal to provide electricity is already the single largest source of increasingly dangerous concentrations of carbon dioxide in the atmosphere, adding to that source will add to the problem—at a time when we desperately need to be turning our climate supertanker around. Besides, America sets the pace for the world. Despite periodic criticism from Europe and infrequent resistance from some developing nations, the world looks to the United States for leadership and the peoples of the world struggle to emulate America. So if America, with one of the highest standards of living on the planet, says that it cannot afford to give up its addiction to coal, there is no reason to hope that any other country—especially any developing country—will struggle to reduce its dependency either.

On the contrary, the International Energy Agency estimated in 2005 that fourteen hundred 1,000-megawatt coal-fired power plants would be built somewhere in the world by 2030. In *Big Coal,* Jeff Goodell describes how that might complicate our collective future: "Right now about one quarter of the world's CO_2 emissions come from coal. If we go ahead with these new coal plants, they will add roughly 570 billion tons of CO_2 to the atmosphere over the life of the plants. (To put that in perspective, 570 billion tons is about as much CO_2 as released by all the coal burned in the last 250 years.) If that happens, our chances of stabilizing the climate are virtually zero." The great thing about all this (if you're a coal company) is that politically, the problem is easily blamed on somebody else. Half of those fourteen hundred proposed coal plants will be built in China, a nation of 1.3 billion people struggling to pull itself out of poverty.

Goodell does a good job of explaining why U.S. coal executives and certain politicians find it so convenient to criticize the world's most populous country: "China's coal habit gives Big Coal supporters in the United States moral cover to argue that taking any meaningful action to limit CO_2 emissions will, as President Bush put it in 2005, 'wreck our economy.' After all, why should America put itself at an economic disadvantage to save the planet if the Chinese won't stop burning coal, too?" Unfortunately, Goodell goes on, the finger of blame points both ways:

> Chinese leaders understand very well that the reason global warming threatens the stability of the planet today is because the industrialization of the West—those 150 blissful years of burning fossil fuels—loaded up the atmosphere with CO_2.

> They also are quite aware that the average American is thirty times richer than the average Chinese, and they don't hesitate to remind people of it. As one Chinese delegate involved in negotiations over the Kyoto Protocol put it, "What [the developed nations] are doing is luxury emissions. What we are doing is survival emissions."

You can see the arguments for inaction piling up on one another. America can't stop burning coal without hurting the already faltering economy. (As Frank Maisano might say, it's just not practical.) But China can't stop burning coal without endangering an already impoverished population. And the sorry performance of each country renders any progressive effort by the other irrelevant. It's like a childish argument about littering: "There's no point in me putting my litter in a trash bin because everyone else is still throwing theirs on the ground." That's tantamount to saying that there is no point in showing leadership or in doing the right thing merely because it's the right thing to do. Both China and the United States are implying that until they can guarantee that everyone else is doing the right thing, they intend to continue being irresponsible, regardless of the path of destruction that sets for the Earth.

The litter analogy is appropriate, because that's what Big Coal is doing: littering. That's what we're all doing when we loft our emissions into the air, whether those wastes come from our cars, from the energy we burn to heat and light our homes, or from the emissions that are created in growing our food and transporting everything we consume. When we put our garbage on the curb, we understand that it costs money for someone to collect that garbage and deal with it safely, and we pay our taxes to cover that cost. But we don't pay the cost of dumping our greenhouse gas garbage into the atmosphere, and we don't ask the world's richest corporations to do so, either.

We have been hearing more and more lately about the coal industry's intentions to stop dumping its garbage skyward. We've been hearing about "carbon capture and storage," in which energy producers capture carbon dioxide in the production process and pump it into depleted oil and gas reservoirs, unmineable coal seams, deep saline formations, or the deep ocean for safekeeping. The theory, as yet mostly untested, is that carbon dioxide pumped and pressurized becomes heavier than air, reducing the likelihood of serious leakage, even if the capping system fails. Proponents theorize further that sequestering carbon in this way would allow us to go on burning all 270 billion tons of American coal—and, for that matter, the entire contents of the Canadian tar sands—without ever having to worry about its effect on climate.

In the current circumstances, however, the whole notion of sequestering carbon is a load of codswallop. In fact that would be the fairest—and the most polite—way to describe the whole notion of "clean coal." Consider this: the Union of Concerned Scientists offers these statistics in reference to what it calls the average 500-megawatt coal plant. On the bright side, this "average" plant produces 3.5 billion kilowatt-hours per year, enough to power a city of about 140,000 people, using 1.4 million tons of coal, 2.2 billion gallons of water, and 146,000 tons of limestone in the process. Then this "clean" coal plant produces the following toxins:

- 10,000 tons of sulfur dioxide, the main component in acid rain;
- 10,200 tons of nitrogen oxide, a major cause of smog and a contributor to acid rain;
- 3.7 million tons of carbon dioxide;
- 500 tons of small particles, a major contributor to lung disease;
- 220 tons of hydrocarbons, smog-producing particles of unburnt fuel;

- 720 tons of carbon monoxide, a greenhouse gas that is also poisonous to humans;
- 125,000 tons of ash and 193,000 tons of sludge from the smokestack scrubber. The ash and sludge consist of coal ash, limestone, and many pollutants, including toxic metals such as lead and mercury. (A coal-ash sludge pond broke through its containment wall in Tennessee in December 2008, knocking a house off its foundation and spreading toxic, heavy-metal-ridden fluid over 400 acres of land); and
- 225 pounds of arsenic, 114 pounds of lead, 4 pounds of cadmium, and many other toxic heavy metals.

There is no question that today's new coal plants are cleaner than those built a century ago. Politicians of both stripes should look back proudly on environmental regulations of the 1970s and the U.S. Clean Air Act of the early 1990s. But just because coal plants are better than before is no justification for an Orwellian redefinition of the word "clean."

That, however, doesn't stop the spinners from trying. Bracewell & Giuliani's Frank Maisano boasted in a December 22, 2008, newsletter that a new coal plant in Florence County, South Carolina, would produce "less than 100 pounds per year" of mercury, which tends to concentrate in fish and then cause birth defects, brain damage, and other ailments in anyone who later eats that fish. For those who are trying to keep track, one hundred pounds is more mercury than you would find in 5 million compact fluorescent lightbulbs. And, of course, the light bulbs can be disposed of responsibly, while the coal plants are spraying mercury into the air.

Once again this list of toxins is the expected output from one 500-megawatt plant (of which the Chinese are currently commissioning two per week). At least half the plants, which in the United States have an average age of thirty-five years, are worse. New plants—those built with the most recent technology that government might demand—burn cleaner than old plants, but that doesn't make any coal clean.

The idea that coal can be burnt cleanly or safely in a warming world, now or soon, is a fiction. The coal companies fight desperately against any effort to restrict toxic emissions, and while they talk a good game on the notion of carbon capture, anyone compelled to speak truthfully agrees that the technology is decades away. *New Scientist* magazine gave a realistic sense of the timeline and anticipated costs in a March 2008 feature titled "Can Coal Live Up to Its Clean Promise?" *New Scientist* reported, "A study by the Massachusetts Institute of Technology called *The Future of Coal,* published last year, suggests that the first commercial CCS [carbon capture and storage] plants won't be on stream until 2030 at the earliest. Thomas Kuhn of the Edison Electric Institute, which represents most U.S. power generators, half of whose fuel is coal, takes a similar line. In September, he told a House Select Committee that commercial deployment of CCS for emissions from large coal-burning power stations will require 25 years of R&D and cost about $20 billion." After a press conference with James Hansen, the director of NASA's Goddard Institute for Space Studies, on June 1, 2007, the Environmental News Service carried a headline that read, "Earth's Climate Approaches a Dangerous Tipping Point." In the copy they went on to describe a "dangerous and irreversible" tipping point, and Hansen did not predict it happening in twenty-five years. He mentioned 2016 but implied that it might come sooner.

The Massachusetts Institute of Technology academics who authored *The Future of Coal* also pointed out that as of the 2007 conclusion of their study, there wasn't a single carbon-capture demonstration project, experimental or otherwise, in any coal-fired facility anywhere in the world. The largest sequestration project of any kind was in the Sleipner gas field in the North Sea, and it was successfully pumping one million tons of carbon dioxide a year into saline aquifers under the ocean—a handsome-sounding total until you consider that U.S. coal-fired utilities alone are currently broadcasting 1,500 times that amount (1.5 billion tons) into the atmosphere every year.

David Ratcliffe, CEO of Southern Company, the second-largest emitter of greenhouse gas from coal in the United States, summarized the challenges of carbon sequestration in an interview on the PBS Frontline documentary *America's Addiction to Coal*. Ratcliffe said, "I think the truth is, is that we don't know where we have storage capability in this nation at this time. We haven't even come close to defining what will be required in storage, what are the legal liabilities and what are the permitting requirements, much less the infrastructure needed to develop that storage and move the carbon, the CO_2, into that storage, pipelines or trucks or whatever that is." Given these hurdles you might wonder where the public got the impression that freshly scrubbed coal is the order of the day or that all new plants are carbon-capture ready or that clean coal is a realistic part of a responsible and healthy energy future.

Well, once again we've been had. We have been Astroturfed by some of the biggest players in the business. As you will read in the coming pages, a reasonable-sounding group called Americans for Balanced Energy Choices spent US$40 million during the last presidential election ensuring that the words "clean coal" were on the lips of every candidate and the screen of every television broadcast for months at a time. These people's idea of clean coal is the same as that of a badly organized college freshman looking through a stinking laundry pile for "clean" underwear. But as James Hansen said two years ago, the current circumstances don't argue for the efficacy of half measures—or for the intelligence of self-delusion. It makes no sense to pursue a self-destructive policy just because it is affordable. Truly clean coal may one day be possible. Carbon sequestration may one day be practical and cost-competitive. But telling Americans that clean coal is burning in their neighborhood power plant today is a fiction that can only do us all a great deal of damage.

"IF PREDICTING THE weather were easy, we'd never be caught in the rain without an umbrella. Predicting weather conditions a day or two in advance is hard enough . . . so just imagine how hard it is to forecast what our climate will be 75 to 100 years in the future." Here you have an example of what the comedian Stephen Colbert calls truthiness. If you don't look too closely at the factual underpinning of that statement, it seems to make perfect sense. It seems true, on the face of it. But it confuses weather with climate and it lays out the supposition that because something is difficult or complicated (anticipating global climate change), there is no reason to try, or at least no reason to listen to the incredibly intelligent people who are doing a Nobel Prize–winning job.

We captured the quote, which seems crafted to make people nod their heads in agreement rather than to actually think seriously about an issue of global importance, in a screen shot from the now-defunct Center for Energy and Economic Development. CEED was a coal-industry front group that arose in 1992 and spent the next fifteen years or more lobbying against new coal-industry regulations and especially against any policy that would rein in greenhouse gas emissions. The CEED Web site also offered this battered bit of denial: "Some scientists believe that the one degree of warming that has taken place over the past 100 years is evidence that potential catastrophic climate change is an imminent threat. However, other *equally-qualified experts* are not so sure. They point to the fact that most of the warming that occurred in the 20th century happened prior to the 1920's—when manmade emissions began a rapid increase" (my emphasis).

In addition to its public and private lobbying activities, CEED carried and promoted the Oregon Petition (which it referred to as the Seitz Petition), urging anyone who could argue their credentials as a "scientist" to sign and support the effort to undermine public faith in climate change science. CEED's leader was president and CEO Stephen L. Miller. In a 2004 memo to Peabody Coal CEO Irl F. Engelhardt outlining CEED's strategy of distraction and delay, Miller wrote:

> In the climate change arena, CEED focuses on three areas:
> - opposing government-mandated controls of greenhouse gases (GHG),
> - opposing "regulation by litigation," and
> - supporting sequestration and technology as the proper vehicles for addressing any reasonable concerns about greenhouse gas concentrations in the atmosphere.

In the memo, now posted at ArizonaEnergy.org, Miller went on to explain that CEED supported sequestration and technology for tactical purposes, saying, "Our belief is that, on climate change like other issues, you must be for something rather than against everything. The combination of carbon sequestration and technology is what we preach and we are looking for more members in the choir." But Miller also makes it clear that preaching about sequestration is different than committing to it: "The other element is to pose voluntary sequestration and technology as the correct policy, rather than mandatory controls." But the industry's track record stands as evidence that a voluntary sequestration policy will result in exactly no sequestration projects. It's hard to believe in that light that Miller is sincere.

The above-referenced screen shots from the CEED Web site featured many pages with these kinds of policy promotions. The site contained numerous descriptions of potential lines of attack on emission regulations and a careful strategy to "sow discord" among states that looked like they might force industry's hand by setting up a regional greenhouse gas initiative.

In its defense, there was never any doubt that CEED was an industry organization. It acknowledged its membership clearly on its Web pages: "Eight Fortune 500 companies are among CEED's nearly 200 member companies and organizations. Over 70 of those companies participate in CEED at the Board Member level. In addition to coal producers, utilities, and railroads, CEED's members also include barge shippers, equipment manufacturers and suppliers, labor unions, and others who know first-hand the economic and social benefits derived from coal-based electricity generation." This clearly is a coalition that can muster significant resources and influence.

Its successor group seems to have been just as well-connected, without being quite so quick to acknowledge its industry connections. Americans for Balanced Energy Choices (ABEC) emerged in 2000, sporting a Web site that, according to SourceWatch.org, was registered to CEED. It's interesting that ABEC's mailing address and fax number were also being used by the Greening Earth Society, the Astroturf group that the Western Fuels Association set up in the early 1990s to argue that climate change would be good for us. Without ever acknowledging its funding partners, the well-financed ABEC set about building "grassroots support" for coal as an appropriate fuel for the provision of electrical energy and to lobby against environmental regulation. In 2002 ABEC launched its first major clean-coal advertising campaign. The resulting television ads, broadcast more than eight hundred times in the pricey D.C. market, proclaimed that "advancements in clean coal technologies are effectively making our environment cleaner" and that "electricity from coal is an increasingly clean source of energy." But Steve Miller and his second in command, Joe Lucas, who were the officers at ABEC, were hardly standing at arm's length from industry. Between 2001 and 2008 the two men reported income in the amount of US$6.4 million for lobbying against stricter emission regulations.

In keeping with the general trend in the fossil fuel community of shifting from denial to delay, Miller and Lucas started to ease away from the harder-edged reputation of both CEED and ABEC and early in 2008 merged the organizations into a new group called the American Coalition for Clean Coal Electricity (ACCCE). By then word had emerged that ABEC had been given a US$35 million war chest for use in fighting climate change regulation. That money was shuffled from ABEC to ACCCE and topped up with another US$5 million, and it helped fund the clean-coal ads that seemed to dominate every CNN commercial break throughout coverage of the November election.

But ACCCE was doing a great deal more than buying air time. Using the Hawthorn Group, a Virginia-based outlet that describes itself as being "among the top 15 independently owned public relations firms in the U.S.," ACCCE began a public and political full-court press, making sure that no prominent political candidate or political reporter could avoid the message that in America, coal is clean—regardless of whether this version of the word accords with the dictionary definition.

The ACCCE tactics were recorded in full in a memo posted online to "Hawthorn Friends and Family." In an attempt to raise clean-coal brand visibility on the ground as well as in paid television slots, the Hawthorn memo said:

- We placed teams in early primary/caucus states, and key battleground states during the fall general election

- We used branding for "clean coal" and "America's Power" consistent with our national advertising campaign
- The team drove a branded, flex-fuel mini-van to events for added visibility
- At each event, we handed out tee shirts and hats with "clean coal" and our logo and Web URL; as well as literature on our issue, to as many event attendees as possible as they stood in line waiting to enter the event
- In the colder months, we also gave out cups of coffee bearing our logo
- Took hundreds of photos and shot video of our activities and posted on our Web site, blog, Facebook page, Flickr account and YouTube channel
- We constantly mobilized our existing grassroots citizen army to mail and email the candidates and ask for support of clean coal technology
- As we attended rallies, campuses, diners and worked town squares, we distributed sign-up cards inviting voters to join our grassroots network
- We routinely emailed our grassroots network our schedule, as well as links to the photos and videos online
- We created and passed out business cards with our Web site, blog, Facebook page, Flickr account and YouTube channel to campaign event attendees.

If you remember seeing a picture of Joe Biden standing with a teenager in a blue clean-coal sweatshirt or Joe the Plumber wearing a white clean-coal baseball cap, it was no accident. Hawthorn and ACCCE made sure those images were transmitted and retransmitted around the country in every form of new media available. And consider for a moment the excitement that must have prevailed in the Hawthorn offices on April 2, 2008, the day that candidate Obama was quoted in the *Scranton Times*

Tribune saying, "And I saw somebody with a clean coal technology hat. We have abundant coal."

The memo's author was Hawthorn's executive vice president and chief communications officer Suzanne Hammelman, and she was exultant in reporting the results of Hawthorn's activities. "We nearly turned candidate events into clean coal rallies," she said, adding that these activities had exactly the desired effect on the electorate: "In September 2007, on the key measurement question—Do you support/oppose the use of coal to generate electricity?—we found 46 percent support and 50 percent oppose. In a 2008 year-end survey that result had shifted to 72 percent support and 22 percent oppose. Not only did we see significantly increased support, opposition was cut by more than half." That, better even than the results of the election itself, tells you who won in the presidential campaign of 2008.

The Hawthorn mission statement is drawn from what the company describes on its HawthornGroup.com home page as Aristotle's definition of "the art of advocacy . . . the ability to find, in any given situation, all the available means of persuasion." But at no point in its efforts to persuade Americans about the cleanliness of coal does Hawthorn report using any actual factual information. There were no safety briefings, no scientific reports, no epidemiological studies of cancer deaths in the lee of giant coal burners. There were bright blue hoodies and sparkling white hats framing the shiny faces of teens and young adults who had been coached to cheer enthusiastically for "their candidate"—which is to say, whatever candidate was in the room. These kids were urged to rush over and stand near politicians and reporters, take one another's photos, and then distribute those pictures online. You got extra points if you could actually get a media photographer or TV camera operator to capture your image instead.

In other words these naïve participants were coached to

game the system—to make fools of anyone who thinks that the practice of politics is still about sincere and committed people standing up for what they believe.

HAVING TURNED THE last dozen pages, you might find it remarkable that your hands haven't turned black. There are undoubtedly scientists and engineers in the coal industry who deserve praise and thanks for their efforts to reduce the pollution that their product creates; there is no question that the technology has improved over time. Neither, however, is there much question about whether industry was at any point eager to implement these improvements.

Arguing that they were just trying to keep rates cheap for consumers, coal-fired utilities have fought regulatory improvements at every turn, and still they advocate for "voluntary" sequestration measures over mandatory controls. This would be so much more convincing if a single company had shown any willingness to volunteer—even to try to create one very small pilot or demonstration project. If they had given US$40 million to university engineering faculty for research instead of spending it on an election advertising campaign—well, we'll never know what the results might have been.

But the Hawthorn memo makes it clear that manipulating American election coverage is both fun and profitable. Forty million dollars might seem like a lot of money, but the coal kings behind ACCCE apparently think it's a small price to pay to influence a new generation of politicians whose in-boxes have been filled with clean-coal emails, and who have grown used to seeing themselves with their arms wrapped around the clean-coal team.

In the land of Washington and Jefferson the maintenance of a healthy democracy is an article of faith. Indeed, much of the world looks to America as an example of democracy in action.

Former president George W. Bush even invoked the spreading of democracy as one of the reasons for starting a war with Iraq. But the Hawthorn campaign, the ACCCE advertising surge, the CEED strategy to disrupt regional climate negotiations among states trying to create a greenhouse gas regulatory system—none of these things scan as truly democratic. They all look like manipulations aimed at taking advantage of a lenient system to privilege the interests of an already wealthy and powerful industry at the expense of the interest of the public. An election campaign is the purest expression of grassroots democracy. This is the occasion on which citizens can make themselves heard, the occasion on which politicians have to present themselves to the people—to listen and to account for their own actions. In 2008 the coal industry stepped into that process in a way that was, by Hawthorn's own description, intentionally disruptive. They played the politicians and the people for fools, and they got the media to go along for the ride. The CEED/ABEC/ACCCE/Hawthorn participants might want to point out that they didn't do anything illegal, but I want to ask if at any point they wondered whether what they were doing was right.

[*fifteen*]

LITTLE COAL

Salvaging a future that's stuck in the tar sands

Canada has had a long history as the United States' principal provider of goods and services. In addition to the longest undefended border in the world, the two countries shared the world's largest trading relationship for decades, and although China bumped Canada as the United States' largest trading partner in 2008, Canada is still the United States' number-one source for raw materials, most notably oil. If you include U.S. imports of both crude oil and refined petroleum, the U.S. Energy Information Administration reports, Canada sends the United States twice as many barrels per day (4.4 million) as Saudi Arabia does (2.2 million).

Of course, Canada sends all manner of high-value human products south of the border as well, including entertainers (from Donald Sutherland and William Shatner to Keanu Reeves and Jim Carrey) and Nobel laureates (such as Rudolph Marcus and Richard Taylor). A slightly less famous Canadian treasure (at least, less famous in the United States) is the environmentalist

Dr. David Suzuki, Canada's answer to Sierra Club founder John Muir. An academic with an environmental conscience and a gift for simplifying science, Suzuki has hosted Canada's most popular TV show on science and the environment, *The Nature of Things,* for three decades, and in 1990 Suzuki and his wife, Tara Cullis, founded the David Suzuki Foundation, Canada's most influential environmental organization. It has been a pleasure to volunteer my services to that foundation almost since its inception, and I am honored today to sit as chair of the David Suzuki Foundation Board of Directors.

One of the best parts of that job is getting to work directly with Suzuki and hear his impressions and interpretations of what's happening in the natural and political worlds. He's a great storyteller, and one of my favorite stories is one he tells about adolescent revelation. When Suzuki was in high school at the London Central Secondary School in Ontario during the early 1950s, he was bright and well-regarded enough that he wound up serving as student council president. But still, he complained at the dinner table one evening that he was finding it hard to make sure that everyone liked him. This is an issue for many teenagers, and Suzuki may have been more prone than most. As Japanese-Canadians, he and his family had been stripped of their possessions and interned during the Second World War. It was a stark example of the price you can pay if you fall out of favor in the neighborhood.

Suzuki's father, however, was obviously unbowed from spending the war at work in Canadian labor camps. He said that if David tried to be liked by everyone, he would never stand for anything. The young Suzuki took that message to heart. During his subsequent career as a geneticist and journalist, he has carried a sense of courage and conviction that I personally have witnessed in few other individuals. After reading Rachel Carson's watershed book *Silent Spring,* Suzuki firmly decided that he stood for environmental awareness, and he has never shied from standing up for it since.

One of David's talents is inspiring those around him to be all that they can be, and from that inspiration, when he recommended me as chair in 2007, I started looking for an opportunity to learn more about corporate governance. I was pleased to right away find a directors education program starting in Calgary, Alberta, through the Institute of Corporate Directors.

I love Calgary. I grew up there, and I go back often to visit family members and friends. But if by his conviction David Suzuki has become a polarizing character in the Canadian establishment, Calgary is the pole where his support is the coldest. It's an oil town, a booming (and busting) community of engineers and entrepreneurs, most of whom make their living from the abundant fossil fuels that have made Alberta the richest jurisdiction in one of the richest countries in the world. People in Calgary don't like to talk about climate change, and these days David Suzuki talks about little else. The directors course, overall a fabulous experience, was also populated largely by oil and gas executives. As a group, they were bright and polite, so while they didn't quite welcome me as one of the family, we quickly forged a working relationship that tended to feature more gentle ribbing than outright argument.

Given that relationship, I wasn't completely surprised when I walked into the room on the morning of February 27, 2007, and found a copy of the *Calgary Herald* spread out in front of my usual place. The large front-page headline said: "Suzuki Says Premier Unfit to Lead Alberta." It seems that the week before, Alberta premier Ed Stelmach had told a business audience that he understood the government had a responsibility to protect the environment, but that "it's clear that green politics are as much about emotion as they are about science." Suzuki the scientist was passing through town on a tour to raise awareness about climate change, and his response included a generous helping of emotion: "If your premier thinks he's worried about the future—and he doesn't realize not doing anything about greenhouse gases is going to wreck the economy—then he doesn't deserve to be a leader." Throughout the day one after another of my director colleagues took the opportunity to stop by my chair or touch my elbow during a break to say, "So, you're David Suzuki's PR guy; was it *your* idea to say that our premier is unfit for office?"

Time and again I smiled and said that David Suzuki, son of Kaoru Suzuki, is not the kind of leader who takes speaking points or tolerates muzzling by his public relations advisor. But as the day wore on, I started wondering more specifically about the quote. I was thinking about David's periodic insistence on saying things that got people riled up and about the circumstances that inspired him to do so on this occasion.

The first thing you might consider is something that the *Herald* reported in the same story: Alberta, with roughly 11 percent of the Canadian population, generates 40 percent of Canada's greenhouse gas emissions. The province's tar sands deposit, the largest source of oil outside of Saudi Arabia, and one of the dirtiest anywhere, is also the single biggest and fastest-growing source of those emissions anywhere in the country. As reported in the overview document *Oil Sands Fever,* prepared by the widely respected Alberta environmental organization the Pembina Institute, the tar sands are huge. Spreading over 58,000 square miles, they cover an area as large as Florida. If Canada is America's biggest (and most reliable) source of oil—and it is—this is where most of that oil originates.

It is also a sticky, toxic mess. As quoted by journalist Andrew Nikiforuk in his book *Tar Sands,* "Dr. Steven Kuznicki, a scholar at the Imperial Oil–Alberta Ingenuity Centre for

Oil Sands Innovation, calls bitumen some of the 'ugliest stuff you ever saw... contaminated, non-homogenous and ill-defined... Bitumen is five percent sulphur, half a percent nitrogen and 1,000 parts per million heavy metals. Its viscosity [stickiness] is like tar on a cold day. That's ugly.'" In order to refine a single barrel of crude oil from this material, Alberta's tar sands plants have to dig up an average of more than two tons of overlay material (in other words, that which used to be the earth and flora of the boreal forest) and two tons of tarry sand. The tar must then be heated using 250 cubic feet of natural gas—enough gas in all tar sands operations to heat three million Canadian homes every year—and washed with between two and five barrels of fresh water, a resource that tar sands companies are licensed to consume at the same daily rate as the entire city of Indianapolis. When you're done, you're left with a vast scar in the Earth and a tailings pond full of toxic sludge.

The tailings ponds are a growing problem. Already covering more than fifty square miles, these are considered to be the largest man-made structures on Earth. NASA's shuttle veterans have reported that you can see the ponds glistening from space, but they are anything but attractive up close. I had the opportunity to fly over this scattered black ocean in a helicopter last year and, even seeing it, the extent of the devastation is hard to digest. The ponds leak an unknown amount of that heavy-metal-laden goop into the surrounding water supply 24/7.[1] One report, prepared by the Pembina Institute in December 2008 for Environmental Defence Canada, estimated that seepage could amount to 3 million gallons per day or a billion gallons a year. Another Environmental Defence document, titled *The Most Destructive Project on Earth,* noted in 2008 that, thanks in part to that seepage, total mercury in sediments in the Athabasca River delta were 98 percent above historic medians, dissolved arsenic in the river had jumped 466 percent, and arsenic in sediment had risen by 114 percent. There is also the question of airborne toxics. A 2007 Environment Canada report noted that summer heat releases huge amounts of volatile organic compounds from the ponds, including over one hundred tons per year of benzene. As with mercury, there is no safe level for exposure to benzene—any amount threatens human health.

The tailings ponds pose a constant risk to wildlife, especially the migratory birds that sometimes make the mistake of landing on what they thought was water. In April 2008 five hundred ducks died an oily death after doing just that. At the time, Syncrude Canada insisted on CTV this was the first such incident in thirty years. That would be easier to believe if Syncrude had actually owned up to the duck deaths, as it is required to do by law. But Syncrude kept the deaths quiet until an anonymous tipster reported the incident to an Alberta Fish and Wildlife office. Syncrude, as well as Suncor, the other major tar sands developer, might also be more credible in defending the state of the ponds if they were not so aggressive in discouraging environmental oversight. Dr. Kevin Timoney described the company's policy this way in a December 9, 2008, article titled "Toll of Oilsands Tailings Ponds on Migratory Birds Is Difficult to Measure" in the *Edmonton Journal:* "Oilsands leases are secured facilities, fenced, patrolled and administered as private property. Studies not supported by industry are not allowed, nor are ornithologists and the general public permitted to observe bird mortality incidents. Few scientific studies have gathered mortality data in a form that allows a real estimate of death rates."

Thus we know only that annual bird deaths are somewhere between sixty-nine, the number reported by industry, and one hundred thousand, the highest estimate offered in a Boreal Songbird Initiative report entitled *Danger in the Nursery.* In the piece he wrote for the *Edmonton Journal,* Timoney, an ecologist with Treeline Ecological Research, dismissed the higher numbers out of hand, suggesting the current annual death toll is in the range of eight hundred to seven thousand. "Such a death toll," he says, "is sufficiently tragic not to require exaggeration."

The tar sands are taking a toll on human health as well, though the details are similarly up for debate. As *National Geographic* reported in a story in March 2009, in 2006 John O'Connor, a family physician who flew in weekly to treat patients in the tar sands community of Fort Chipewyan, reported that he had in a short period seen five cases of cholangiocarcinoma—a bile-duct cancer that normally strikes one person in one hundred thousand. There are only one thousand people total in Fort Chip. O'Connor couldn't seem to interest anyone in the cancer cluster until he mentioned it in a 2006 radio interview. Then, he told *National Geographic* writer Robert Kunzig, the story "just exploded." But it still didn't interest the government. Alberta Health spokesperson Howard May responded in an email to Kunzig, "There is no evidence of elevated cancer rates in the community." It's not clear how May justified making this statement when O'Connor's statistics proved that the occurrence rate of cholangiocarcinoma in Fort Chip was five thousand times above normal.

How is government responding to these risks? Well, instead of stepping up the monitoring of local health or the enforcement of environmental standards, Alberta premier Ed Stelmach added C$25 million to his 2008 budget for a public relations campaign aimed at promoting Alberta and the tar sands as environmentally friendly.[2] This is a classic mistake—a misstep that should send off warning bells to anyone with experience in public relations. The first priority in crisis management should always be to manage the crisis, not the media. If ducks die, you should think about ways to prevent them from dying in the future. If people die, you should apologize to their family members and try to identify and prevent the cause of death. If your province is growing incredibly rich at the expense of other people in the world, you should look hard at whether it is absolutely necessary to authorize the industry responsible to double in size over the next five to seven years.

That's the current Stelmach plan: double the tar sands operation as quickly as possible. As *National Geographic* reported, oil sands developers spent C$50 billion in construction between 1998 and 2008—C$20 billion in 2008 alone. And the plan, at least before oil prices tanked, was to spend another C$100 billion to more than double the tar sands output between now and 2015. You can imagine the pressure that must exert on a premier, a governor-equivalent reigning over a province (Alberta is 255,000 square miles) that is only a little smaller than Texas (268,000 square miles) but has a population (3.6 million) that is only a little larger than that of Connecticut (3.5 million). You can imagine that Premier Stelmach would think it politically untenable to turn off, or even slow down, that kind of development.

But that depends on whom he might consider to be his political masters. In May 2007 a Pembina Institute poll found that 71 percent of Alberta respondents were in favor of a moratorium on new tar sands development. So it wasn't Stelmach's voters who were urging him to overheat the economy and the world (the mining and refining of tar sands oil generates three times as much greenhouse gas as conventional oil); it was a different stakeholder group altogether.

Faced with the prospect of placing a moratorium on tar sand development, even only until some of the safety issues could be worked through more effectively, Stelmach responded, as reported in *National Geographic,* like this: "It's my belief that when government attempts to manipulate the free market, bad things happen. The free-market system will solve this." But as *National Geographic* writer Kunzig pointed out, the free market doesn't consider the effects of tar sands pollution on rivers and forests, cancer-stricken neighbors, or people around the world at risk from greenhouse gas emissions. Kunzig concludes, "In northern Alberta, the question of how to strike that balance has been left to the free market, and its answer has been to forget about tomorrow. Tomorrow is not its job."

Alberta's oil producers are not lifeguards. They are business-people, bound by the rules of business to maximize profits for their shareholders. And it doesn't appear that Alberta's government is overburdened with lifeguards either. Without getting into a debate over Premier Stelmach's fitness to govern, he seems disturbingly willing to stand back and let people stampede toward the cliff. As the government says in its January 2008 climate change plan, *Responsibility/Leadership/Action,* "As a leading energy producer for Canada and the world, Alberta is responsible for producing about a third of Canada's total greenhouse gas emissions—the leading cause of climate change. Our emissions are expected to increase by another third over the next five to ten years." Tomorrow, apparently, is not the Alberta government's job either.

That, unfortunately, leaves everything up to you. No matter where you live, whether you're an Albertan in Premier Stelmach's own provincial riding or a Californian filling up your car with tar sands gasoline, you're the person who is going to have to start taking responsibility, not just for your own actions, but for the position of your government and the integrity of the public climate change conversation. That's a daunting prospect, for reasons that I will explore at greater length in the next chapter. It's clear that the political and economic momentum is pushing us hard and at ever-increasing speed in the wrong direction. But it's just as clear that personal action by each of us will be essential to changing course because in this jurisdiction, at this time, the lifeguards are off duty.

[*sixteen*]

NOBODY WANTS TO BE A CHUMP

How the debate cripples public policy and paralyzes private action

In a number of the great powers, climate change scenarios are already playing a large and increasing role in the military planning process. Rationally, you would expect this to be the case, because each country pays its professional military establishment to identify and counter "threats" to security, but the implications of their scenarios are still alarming. There is a significant probability of wars, including even nuclear wars, if we ever reach the range of +2 to +3 degrees Celsius [4 to 5 degrees Fahrenheit] hotter. Once that happens, all hope of international cooperation to curb emissions and stop the warming goes out the window.

The activists who warn about the consequences of climate change generally do not go too deeply into these issues, perhaps for fear of inducing despair in those whom they seek to mobilize. Back in the days when the climate change denial industry was still manufacturing a fake "debate" to cast doubt on the whole phenomenon of global warming, especially in

> North America, there may have been a tactical case for soft-pedalling the consequences for fear of sounding too extreme. (Although I don't think so really. It almost always turns out to be better in the end to state the facts as you see them as clearly as possible, and let the chips fall where they may.) At any rate, the denial industry is now in full retreat, and it's high time for everybody to say exactly what they mean.

That is the flyleaf version of an introduction to Gwynne Dyer's excellent and sometimes apocalyptic book *Climate Wars*. Dyer is no agenda-driven alarmist, even if he is raising the alarm. He's a former sailor, having served in the Canadian, American, and British navies, who went back to school and did a Ph.D. in war studies at the University of London, in the United Kingdom. He has taught at the Canadian Forces College and at Britain's Royal Military Academy Sandhurst. He is also a respected geopolitical analyst whose writings appear regularly in 175 newspapers and whose broadcast efforts, culminating in the celebrated NFB-CBC series *War*, have won him accolades that include an Academy Award nomination.

Dyer stumbled onto the topic of global warming all but inadvertently. A couple of years ago he noticed an outbreak of reports from senior military analysts and defense-industry think tanks looking at the risks and dangers in a climate-changed world. Immersing himself in the field, Dyer grew increasingly nervous, partly because the scenarios were so bleak and partly because the conversation about alternatives was stifled by what he refers to as "the Bush ban on treating climate change as a real and serious problem."

Thus in the United States, we have had a situation in which a government, under pressure from its biggest and most self-interested supporters, actively frustrated the efforts of its own military to protect the country from risk. We also have seen an attack on the integrity of the scientific community and a disinformation campaign that was so well-funded and widely executed that the public is lost in confusion. Businesses, individuals, and governments in North America have, as of this writing, done nothing to even constrain the growth of greenhouse gas emissions, much less begin to meet agreed targets for reduction.

But Dyer comes off the rails in his conclusion that "the denial industry is now in full retreat." On the contrary, the evidence indicates that the denial industry is digging in with more determination than ever, even if it is growing more sophisticated in its tactics.

The resistance was perhaps predictable. The great Chinese military strategist Sun Tzu had this figured out in the 6th century BCE when he wrote his definitive manual, *The Art of War*. He wrote: "When you surround an army, leave an outlet free. Do not press a desperate foe too hard." The reason, Sun Tzu said, is that soldiers who find they have no room for escape "will prefer death to flight . . . Officers and men alike will put forth their uttermost strength." That's a reasonable description of the position the deniers and their sponsors find themselves in today. The Big Oil and Big Coal barons, and the public relations advisors, lobbyists, and junk scientists who have been advancing their case, have no reasonable line of retreat. ExxonMobil, which for the past four or five years has been recording bigger profits than any company in the history of companies, has invested almost nothing in renewable or alternative energy sources and next to nothing in climate change mitigation strategies such as carbon sequestration. The coal, railroad, and electrical utility companies have similarly committed themselves to a strategy of disinformation and resistance rather than working diligently to figure out ways to use coal safely or use alternative sources of energy.

As for die-hard deniers like Fred Singer or Steven Milloy, it would be incredibly difficult for them today to accept, in a modern epiphany, the undeniability of climate science. Rather than retreating, they have redoubled their efforts. The Heartland Institute held the first "International Conference on Climate Change" in 2008, the same year that Dyer published *Climate Wars,* and they advertised the 2009 conference as "the world's largest-ever gathering of global warming skeptics." The creators of the Oregon Petition, as well, continue to troll for signatures, having plumped their suspicious numbers from seventeen thousand skeptical "scientists" in 1999 to more than thirty thousand in 2009.

The DeSmogBlog's Kevin Grandia did a quick analysis for the Huffington Post in January 2009 on the frequency of denier material being posted on the Internet and found that skeptical content had at least doubled from the previous year. Searching for linked phrases such as "global warming" with "hoax," "global warming" with "lie," and "global warming" with "alarmist," Grandia found that far from being in retreat, the "officers and men" of the denial industry were fighting like never before.

Gwynne Dyer, it seems, has fallen into a trap common to people who do a lot of research on climate science and its implications. As they immerse themselves in the serious science, they become chillingly certain about the threat, and they begin to dismiss the notion that anyone could still be arguing the point. As mentioned in Chapter 9, Ross Gelbspan told me in 2005 that the denial campaign was kaput—only to discover a few short months later that it was more active than ever.

The evidence of the denial movement's success is also everywhere around us, but nowhere near as clear as when you ask people whether they believe there is still a scientific dispute. In September and October of 2008 researchers from the Yale Project on Climate Change and the George Mason University Center for Climate Change Communication surveyed more than 2,100 Americans on their knowledge and opinions about climate change. In the resulting report, called *Climate*

Change in the American Mind, 63 percent of respondents said that they were personally worried about climate change. But when asked whether "most scientists think global warming is happening," just 47 percent said yes; 33 percent said "there is a lot of disagreement" among scientists; 3 percent said "most scientists think global warming is not happening;" and 18 percent just didn't know. Looking back at Naomi Oreskes's *Science* article "Beyond the Ivory Tower: The Scientific Consensus about Climate Change," which recorded virtual unanimity in the scientific community, you realize that the continuing popular belief in the existence of a scientific debate can only stand as a tribute to the success of the disinformation effort.

The upshot of that success is twofold. First and most obviously, it has prevented, in North America at least, the implementation of any serious policy initiatives to mitigate climate change. People who believe scientists may still be arguing over the details are not going to demand urgent action from their government, and—quite clearly—governments that are in thrall of (or in debt to) private interests that are profiting from inaction are not about to impose policies that will discomfit their voters and offend their financial backers.

But what Al Gore described in the title of his 2008 book as the "assault on reason" has done more than prevent governments from acting to protect us all from the effects of global warming. The constant campaigns by organizations like TASSC—the attack on scientific integrity and literal truth that has been financed by industry players such as Philip Morris and ExxonMobil—has served to undermine public trust in a way that may actually endanger the democratic process.

In the months following the September 2008 meltdown of the global banking system it wasn't difficult to find evidence that many people were losing faith. The Canadian Association of Petroleum Producers published research in January 2009 showing that 50 percent of Canadians "do not believe what oil and gas executives say in the media," compared to just 13 percent who do. The international public relations firm Edelman reported in March 2009 that its 10th annual "Trust Barometer" opinion poll had found that public faith in industry had hit an unprecedented low. After interviewing more than 4,400 people in 20 countries, Edelman found that in the United States, for example, 77 percent of the people trusted business less than they had just one year earlier.

In *Climate Change and the American Mind,* the Yale/George Mason University team also asked whom Americans trust as a source of information about global warming. It found that 82 percent of Americans trusted scientists, followed by environmental organizations (66 percent) and television weather reporters (66 percent). About half (47 percent) of Americans trusted the mainstream news media. Only 19 percent of Americans trusted corporations as a source of information. Against those results, which mirror Canadian results from a 2007 survey by Angus Reid Strategies, it seems that the Canadian Association of Petroleum Producers may be overestimating their credibility in the public eye.

When I first saw the Angus Reid results in 2007, I was both surprised and concerned. As chair of the David Suzuki Foundation, I am gratified that environmental organizations have credibility. But that only covers one of my volunteer commitments. As the owner of a public relations company whose work comes mostly from corporations, I began to wonder, if the public doesn't trust corporations, what do they think about public relations people? So for the next Angus Reid omnibus poll, I asked them to attach a question. They asked:

> Which of the following statements best represents your own opinion of the role and function performed by public relations experts?

- PR experts help the public better understand the environmental performance of companies by providing clear and accurate information.
- PR experts help deceive the public by making the environmental performance of companies appear better than it really is.

I was expecting a negative answer, but I wasn't expecting to hear it from so many people: 81 percent of respondents said that public relations experts help deceive the public. This suggests to me that the spin campaigns of the last century have taken their toll. Those campaigns, beginning with Edward Bernays and the smokin' debutants in the Torches of Liberty parade and culminating with the massive campaign to deny what may be the greatest environmental threat ever to face humankind, may have succeeded in their short-term goals of fooling consumers and voters into doing the wrong thing. But the worst performers in the public relations world have also critically undermined public faith in business. They also seem to have made some members of the public hopelessly cynical about government and about the quality or reliability of the public conversation.

That, I fear, is why voters stay home. The turnout in U.S. midterm congressional elections has dropped to less than 40 percent. And fewer than 60 percent of Canadians participated in the 2008 federal election. If people don't trust their leaders, they lose faith in the value of their own vote. And why wouldn't they? According to research that the University of Chicago compiled over the last decade, 71 percent of Americans have consistently supported a government-mandated increase in fuel-efficiency standards—during which time Congress has consistently supported the Big Three automakers against the interests of the American people and the planet.

There is yet another element of mistrust, and this one might be the worst of the lot. In 2006 Hoggan & Associates led a research project on Canadian public perceptions and understanding of the concept of sustainability. In concert with a blue-ribbon group of participants, ranging from major energy and transportation firms to local and provincial governments and universities, we initiated a multilevel investigation into what people think of sustainability, asking everything from whether they understand the term to why they do (or do not) incorporate sustainable behavior into their daily lives.

The answer to the last question proved interesting. When we asked people why they don't act more sustainably, they offered a broad and reasonable set of barriers: 45 percent blamed lack of government leadership; 43 percent said they didn't know enough about solutions; 32 percent complained about poorly designed cities and workplaces; and 31 percent said they felt helpless to solve the problem alone. Only 5 percent said they didn't care. But when we asked those same respondents why they thought their neighbors didn't act sustainably, we got a surprising response. Fully 50 percent said they thought their neighbors were "not really concerned."

Together, all these polls seem to indicate the following: people don't trust business; they don't trust government; and on issues of sustainability at least, half the people don't even trust one another. No wonder so few people are struggling to make a large personal contribution in the battle to limit the effects of climate change: nobody wants to be a chump. Nobody wants to be the only person on the block who is spending money to repower their heating system. No one wants to give up their car, change their diet, or limit their consumption if their efforts will be rendered irrelevant by the consumption patterns of those around them.

That leaves us in a grim space. We don't trust our neighbors to do the right thing. We don't trust our leaders to look after our interests. And we doubt our personal ability to stop climate change that seems both debatable and increasingly unavoidable. But before you conclude that it's best to just get on with your life and hope this whole thing will fade away, Gwynne Dyer, writing in *Climate Wars,* offers the following as one of the possible future scenarios generated by the Center for a New American Security:

> And so we are plunged into the nightmare world of scenario two, a world only thirty years hence in which the average global surface temperature is 2.6 degrees Celsius [4.7 degrees Fahrenheit] above 1990 levels, with higher temperatures over land and much higher temperatures in the high latitudes. Accelerated melting of the Greenland and Antarctic Ice Sheets has already raised sea levels worldwide by half a meter [twenty inches], and storm surges driven by much more powerful weather systems are already causing crippling inundations in low-lying port cities like New York, Rotterdam, Bombay and Shanghai. London might buy itself fifty or a hundred years by building a second, higher Thames Barrier, but in general the outlook is for successive retreats inland to new, makeshift ports that will eventually be inundated in their turn as the sea level continues to rise. This continuing abandonment of existing assets and reinvestment in new, temporary port facilities will impose heavy burdens even on once-rich societies.
>
> Meanwhile, densely populated river deltas such as those in Bangladesh, Egypt and Vietnam are already generating huge numbers of refugees as the land is eaten away by successive storm surges. Crop yields are falling steeply in these regions (which provide a disproportionate amount of the world's food). The irreversible destabilization of the ice sheets means that a further sea-level rise of four to six meters [thirteen to twenty feet] is inevitable over the next few centuries,

> so all the major river deltas are ultimately doomed, and civilization is condemned to centuries of continuous retreat as coastal lands are drowned.
>
> Agriculture has become "essentially non-viable" in the dry subtropics as "irrigation becomes exceptionally difficult because of dwindling water supplies, and soil salination is exacerbated by more rapid evaporation of water from irrigated fields." Desertification is spreading in the lower mid-latitudes. Fisheries are damaged worldwide by coral bleaching, ocean acidification, and the substantial loss of coastal nursery wetlands—but then most major ocean fisheries will probably have collapsed through overfishing well before 2040 anyway.

In other parts of the book Dyer talks more about what all this flooding and storming might mean. He suggests, for example, that when "agriculture . . . become[s] 'essentially non-viable' in the dry subtropics," people in nuclear-armed countries like India and Pakistan might start to take that personally. You begin to understand why Dyer says that activists avoid talking about consequences for fear of pushing people directly from denial to despair. It's true that this particular scenario is particularly dire—it sits at the outer edge of what is expected in terms of rising temperatures. But political pressure has forced the IPCC to work only with the most well-established (and therefore older) data, while the observed world climate has been shifting with unprecedented speed. Just as an example, the IPCC Fourth Assessment Report suggests "Arctic late-summer sea ice disappears almost entirely by the latter part of the 21st century." But after two successive years with the lowest ice cover since satellite monitoring of Arctic ice began in 1979, the National Snow and Ice Data Center in Boulder, Colorado, is now suggesting that the late-summer sea ice could be clear as early as 2013—not seventy or ninety years from now, but four or five. Thus, Dyer's

Center for a New American Society scenario is worryingly within the bounds of what might happen, especially if we continue to ignore the problem.

Yet our industry and government leaders are urging us to do just that: ignore the problem. When someone says it isn't practical to address climate change, when they say there is no reason for countries like the United States and Canada to show leadership when China and India may not fall immediately into step, the deniers and delayers are urging us ever closer to the cliff without regard for who or how many people might fall to the rocks below.

In that light we actually face two challenges right now, one less daunting than the other. The easier test, if we chose to grapple with it, would be to create the technology and impose the discipline necessary to deal with climate change. Indeed, a June 2008 report by the McKinsey Global Institute suggests that the climate crisis is manageable, and that the costs are manageable as well. McKinsey estimates that the macroeconomic costs of what it calls the "carbon revolution" would be between 0.6 and 1.4 percent of GDP by 2030. McKinsey adds, "To put this figure in perspective, if one were to view this spending as a form of insurance against potential damage due to climate change, it might be relevant to compare it to global spending on insurance, which was 3.3 percent of GDP in 2005." So mitigating the threats of climate change is the easier part: take out the insurance and minimize the risk. You do that on your house every year; the bank insists. In most jurisdictions, the government has another set of laws that force you to insure your car as well. And in both cases the risk of disaster is significantly less than the greater than 90 percent certainty that scientists ascribe to the climate crisis.

The more difficult task will be to forge the political consensus nationally and internationally to face this problem head on. For the world is not primarily wrestling at this moment with a climate crisis—that's proceeding apace with very little engagement. The world is in the grip of a crisis of political inaction, a crisis of political leadership—or a complete lack thereof.

None of this is to suggest that mitigating the threats of climate change will in fact be easy. When McKinsey talks about a carbon revolution, he strikes the right tone. Look at any of the good scientific books and reports on this subject and they will tell you that we need to reduce our carbon dioxide output by something close to 80 percent by 2030 to be comfortably assured of avoiding the Center for a New American Security's disaster scenario. That doesn't mean that we can continue building coal-fired power plants (carbon-capture ready *after* 2018) as fast as possible or stay the course to triple tar sands production in Alberta, "stabilizing emissions," as Canadian prime minister Stephen Harper proposes, only in 2020. That means we will have to make every effort to conserve energy, to constrain increases in large carbon sources, and to invest heavily in energy alternatives.

To get to those goals, as *New York Times* writer Thomas Friedman likes to say, it will not be enough to change our lightbulbs. We will have to change our leaders. We surely will have to change our own behavior, as a necessary first step and as a show of good personal leadership—of public good faith so your neighbors know that they can trust *you*. Then we have to start demanding more from our leaders—in business and in government. The argument Friedman always makes is that we can change the lightbulbs in our house, but Congress can pass a law that says *everybody* has to change. And that's the pace of change we need. We need, all of us, to get on course for the last chapter.

[*seventeen*]

SAVING THE WORLD

Tactics for turning back the clock on global disaster

Here's the problem: the current public policy on the question of climate change is based on a lie—a carefully constructed, aggressively disseminated lie. In saying so, and before everyone named in this book starts dialling up their libel lawyers, I want to make something clear. I am not referring to the first definition in my dog-eared 1996 edition of the *Oxford English Reference Dictionary,* "an intentionally false statement *(tell a lie).*" Rather, or more frequently, I am thinking of a variant within the second definition, "imposture; false belief *(live a lie).*" We are all living a lie, all ambling along as if everything is going to turn out fine with the climate and the future, even if we currently find it "impractical" to make any realistic gesture to ensure that that is so.

I also want to be clear that in invoking the campaign of misinformation and in complaining about the lack of leadership, I am not, for example, accusing Alberta premier Ed Stelmach of being a liar. Neither am I suggesting any such thing about the oil executives in my directors course, the men and women who gave me such a hard time about the David Suzuki quote. These people live in a world in which acknowledging the reality and the dangers of climate change is an extraordinarily difficult thing to do. It's difficult because facing the true threat of the climate crisis flies in the face of their own self-interest. In making this point in his public presentations, Al Gore often raises a favorite quotation from the American novelist and politician Upton Sinclair, who said, "It's difficult to get a man to understand something when his job depends on him not understanding it."

It's also difficult for your average rising oil executive to take a bold environmental position on climate change—to speak, or even know, the truth—because very few people in the executive offices in Calgary or Houston are likely to stumble accidentally across unvarnished science. If you rely for most of your information on the reports, blogs, and newsletters of the oil patch, or if you confine yourself to newspapers like the *Calgary Herald,* the *Wall Street Journal,* or Canada's *National Post,* you will be consuming a steady diet of stories that, even today, suggest that some aspects of climate science are still in doubt or that overcoming the challenges will be so difficult and expensive that in the short term, we should maintain the status quo.

I am sympathetic to my director colleagues, but I'm not at all forgiving of political leaders like Oklahoma Senator James Inhofe, who has so successfully avoided reading anything that might clear away his aggressive misinterpretations of climate science. I'm not forgiving of Alberta premier Ed Stelmach. He, like Senator Inhofe, has a responsibility, not to the oil companies who pay for his campaigns, but to the citizens he represents, people who deserve to have their interests—and their lives—protected from oil companies that can add US$46.6 million in a single year to the US$82 million that they already spend trying to influence politicians on Capitol Hill.

We have seen the power of that kind of money. We have in the preceding pages seen both the ruthlessness and the resources that can be brought to bear. In a world where mainstream media are overwhelmed, and sometimes caught napping, a well-financed campaigner can recast reality or redefine the character of a political opponent. Consider the 2004 presidential campaign, in which Marc Morano and company attacked the good reputation of John Kerry. There is no question that Kerry is a war hero. There is documentary evidence, and witnesses will stand by his side, decades after the fact, and attest to his courage and selflessness on the rivers and in the jungles of Vietnam. Yet Morano's compatriots were able to marshal the so-called Swift Boat Veterans for Truth and, for long enough to affect a presidential vote, throw Kerry's good reputation into doubt.

The people who want to continue burning coal, selling oil, and mining tar sands have been equally effective. They have told us that their resources are the only ones that will run our economies affordably, and they have ridiculed environmentalists as agenda-driven loonies—"chicken littles" who scream nervously about a sky that is getting oppressively heavy. Sometimes, the most aggressive people in environmental organizations have contributed to that image. Sometimes in moments of frustration or desperation, they have chained themselves to trees or smashed their ships into whaling vessels, adding to the image of environmentalists as inherently radical.

That tide is turning. Go to any event featuring Al Gore or David Suzuki today and you will see a crowd much bigger and much less apologetic than what you might have seen ten years ago. There is a gathering community of leaders—people like General Electric CEO Jeffrey Immelt, Virgin brand owner Sir Richard Branson, and Interface Global CEO Ray Anderson—who have come to understand the problem and who are refusing to let the lie linger any longer.

But they need your help. Susan Nall Bales, president of the FrameWorks Institute, always says that the media set the public policy agenda, and if the media are timid or are being manipulated, the agenda will slip into the hands of those with the most power or the best strategy for affecting the public debate. It can take a long time in that environment for truth to emerge of its own accord, and climate change is not an issue that leaves us a huge budget of time.

So what to do? First and most critically, you must inform yourself. My best advice might be that you should survey a variety of sources just to help confirm—or challenge—what you have read in this book. I am confident that it will stand up to scrutiny, but I am even more concerned that you be rock solid in your own understanding, in your conviction, of what has been happening in the global climate change conversation.

Then, if you haven't done so already, you should read up on climate change science. Read the summary of the latest IPCC report. Read one of the books I mentioned in Chapter 2, or just google *New Yorker* writer Elizabeth Kolbert's Climate of Man series. There is enough information there to whet your appetite, perhaps even to upset your sleep.

While you are reading, you should be hypervigilant about sources. I made a passing reference in Chapter 11 to the rules of evidence that prevail in North American courtrooms. I have a law degree from the University of Victoria in British Columbia, and while I have never practiced, I know some of the conventions. One is that experts in a court case have to be "qualified" by the court. The lawyer who wants to use the expert must first submit to the judge or the jury the expert's credentials. You can't just drag someone off the street and call him a climate scientist.

That's the standard that should prevail in the public conversation, but given the job currently being done by many reporters, you have to take it upon yourself to "qualify" experts yourself. When I hear someone holding forth, I always ask myself these three questions:

1. Does this "expert" have relevant credentials? For example, have they trained in an area of science that is at the very least connected to climatology or atmospheric physics?
2. If an "expert" is talking about science, are they still practicing science? Are they still conducting research and publishing in legitimate peer-reviewed journals? Or are all of their "scientific" pronouncements appearing on newspaper opinion pages, edited by people who think it's just great to provoke debate?
3. Is this "expert" taking money from vested interests or is he or she associated with ideological think tanks—the people who rely for their employment on promoting the agenda of their major funders?

Charming and articulate voices abound out there. People like Bjørn Lomborg seem sincere, and their arguments would make it easier to remain complacent—to doubt the certainty of science or, in Lomborg's case, to neglect action on the basis that some other, equally neglected, priority is higher. But this is not a time for easy answers. This is a time for right answers, which you will find only if you insist on the best sources, the respected journals and national science academies that have no agenda other than advancing the scope of human knowledge.

I must warn you that reading very much of this material can be incredibly depressing. As in the scenarios from Gwynne Dyer's dire *Climate Wars,* a future in which we fail to address climate change includes death, disruption, extinction, and suffering on a massive scale. It's horrible to consider. I believe that's one of the reasons that lie has survived so long: few people really want to sit down and contemplate that dark future. But I wager that if you dig further into the literature about climate change, you will come away with renewed vigor and a righteous sense of justifiable anger at those who have manipulated the climate conversation to date. You will join the neighborhood watch of people who will no longer stand for disinformation to be passed around your social circle—or to go unchallenged when uttered by your local politicians.

That's what we need: vigilance. Eyes on the street. Actually, we need feet on the street, and in large numbers. We need crowds of people demanding that politicians face this issue directly and sincerely. As *New York Times* writer Thomas Friedman argues in his book *Hot, Flat, and Crowded,* we need a social change force on the scale of the civil rights movement—and given the urgency of the climate change threat, we need it to be running at ten times the speed.

We need leadership. That doesn't mean enabling today's "leaders" to keep their positions and expand their power. It means searching out and supporting people who understand that true leadership involves faithfully representing the interests of the people who have asked you to lead. True leaders don't spend their time in self-promotion, advancing their own position at the expense of those around them. True leaders create an opportunity for everyone to make a greater contribution—and to enjoy a greater benefit. We will need leaders like California governor Arnold Schwarzenegger, who stood up to the auto industry in an effort to reduce fleet emissions. We need leaders like British Columbia premier Gordon Campbell, who risked the wrath of voters by passing the first carbon tax in North America. And if those leaders are to survive—if President Barack Obama, for example, is to live up to the expectations of his most optimistic supporters—they will need our support, on election day and every day.

I think again about David Suzuki and the leadership he has shown, and his periodic impatience that has sometimes stirred controversy. A year or so after he got caught dismissing the leadership credentials of the Alberta premier, David was in Montreal speaking to a group of students at McGill University. He was quoted in the *McGill Daily* as saying, "What I would challenge you to do is to put a lot of effort into trying to see whether there's a legal way of throwing our so-called leaders into jail because what they're doing is a criminal act."

I spoke to David afterwards, and he suggested that the writer might have lost some of the subtleties of his point while wrestling his words onto the page, but I suspect the quote was largely correct. It was also interesting that in the first mainstream newspaper report of the story, Craig Offman at the *National Post* actually investigated the literal possibility of jailing our leaders, finding that in Canada, for example, the federal cabinet was in violation of a Canadian law requiring the government to abide by the immediate-term commitments of the Kyoto Protocol.

I don't think that Prime Minister Stephen Harper is going to punish any of his ministers or volunteer for jail time himself, but it got me thinking again about what must be going through David Suzuki's mind. He has dedicated his life to searching out and sharing knowledge about the environment and the prospect of human sustainability. It's all he thinks about. He brings to bear an incredibly sharp mind, a specific expertise in genetics, and a prodigious work ethic. He's always reading.

I am therefore not the least surprised that he is outraged, that he periodically snarls. I am angry enough, and I haven't read a fraction of the material that David regularly consumes on science and the environment. So at the end of the day, I come back to the question of whether, as "David Suzuki's public relations advisor," I would ever have told him to say that a premier was unfit for office or that a governor should go to jail. The short answer, probably, is, "no." I wouldn't have told him to say it.

But the more I think about it, the more I think that he's right. (And I must stress that this is not—on my part or David's—a reflection of David Suzuki Foundation policy.) If our current politicians won't take responsibility for dealing with climate change, then we have to find some who will. If governors and senators will not hold the coal companies to account, then we must cast off the blue T-shirts and the white hats and start demanding truly clean coal—or no coal. If oil companies insist that they need to invade ocean bottoms and nature preserves in search of new supplies, even while they make little or no investment in researching alternative energy, we must reject their demands out of hand. If municipal politicians want to spend money on roads and bridges rather than buses and bike lanes, we have to start writing letters and showing up at meetings—or at the very least supporting those who we *know* are advocating for the right things. And if fossil fuel–funded think tanks are paying people to phone opinion-page editors and flatter them each time they run an article suggesting that climate change is a ruse, then we must phone at the same time and offer a more appropriate critique and, if necessary, a cancellation of our subscription. We need to wrest the public policy agenda away from those who are pursuing self-interest, and return to the notion of public interest.

We have to get informed, and we have to get active. Because if we don't, if we don't all take the initiative and demand of our leaders that they start fixing this problem, beginning today, it will indeed be a crime. And the punishment will be visited on our children and on their children through a world that is unrecognizable, perhaps uninhabitable.

If you are not near a window, a park, or a garden, I encourage you to go find one. Stand under a big tree or contemplate a small flower. A fragile beauty and a tenuous balance both exist in the natural world. Scientists such as NASA's James Hansen tell us that we have upset that balance. They also suggest that we might still undo what we have done. They say that in fifty years our children or theirs may still enjoy the same variety of bird-song and butterflies. There may *not* be nuclear devastation on the plains of India or in the mountains of Pakistan. The oceans may still lap the shores of Bangladesh and Holland rather than coursing through the streets, dispatching desperate refugees to neighboring countries already in crisis.

There can be a good future if we make it so. But if we stand about, if we allow energy-industry flunkies to control the conversation—or even if we just let it ride, cynically accepting that politics is inherently corrupt and that nothing we do can make a difference—we will all have time to regret the passing of a beautiful and sustainable world.

So please, be bold. Be courageous. Be positive. Act and demand action. This, for our sake and for the sake of all those who follow, is a fight that we can and must win. For this bears repeating: the world is worth saving.

NOTES

CHAPTER 2

1 Jeff Goodell, *Big Coal* (Houghton Mifflin, 2006), p. 179.

CHAPTER 4

1 Naomi Oreskes, "You CAN Argue with the Facts," lecture (Smart Energy, 2008), http://smartenergyshow.com/node/67.
2 TobaccoFreedom.org, "Secret Documents Reveal A.C.L.U. Tobacco Industry Ties," www.tobaccofreedom.org/issues/documents/aclu.
3 John C. Stauber, "Smokers' Hacks: the Tobacco Lobby's PR Front Groups," PR *Watch*, volume 1, no 3 (1994), www.prwatch.org/prwissues/1994Q3/hacks.html.
4 APCO Associates, "Revised Plan for the Public Launching of TASSC (Through 1993)", http://tobaccodocuments.org/pm/2045930493-0504.html.
5 Memorandum from Tom Hockaday and Neal Cohen to Matt Winokur of APCO Associates, March 25, 1994, http://tobaccodocuments.org/pm/2024233595-3602.html.
6 See note 4.
7 Memo from Alexander Holtzman to Bill Murray of Philip Morris Companies Inc., August 31, 1989, http://tobaccodocuments.org/pm/2023266534.html.
8 Sheldon Rampton and John Stauber, "How Big Tobacco Helped Create the 'Junkman,'" PR *Watch*, volume 7, no 3 (2000), www.prwatch.org/prwissues/2000Q3/junkman.html.

9 Greenpeace International, "Denial and Deception: A Chronicle of ExxonMobil's Efforts to Corrupt the Debate on Global Warming," www.greenpeace.org/usa/assets/binaries/leaked-api-comms-plan-1998 (or, for a more readable version, see www.euronet.nl/users/e_wesker/ew@shell/API-prop.html).

CHAPTER 5

1 Friends of Science membership quarterly newsletter, No. 7 (July 2005), http://sourcewatch.org/images/2/22/FOS_2005_Q2_Newsletter_7.pdf.
2 "Special Investigations Research and Trust Accounts RT73-2028 &TR73-2153," University of Calgary Audit, April 1, 2008, www.desmogblog.com/sites/beta.desmogblog.com/files/Auditor%27s%20Report%20April%2014%202008.pdf.
3 Natural Resources Stewardship Project, NRSP Background, http://nrsp.com/background.html.
4 Podcast of Dr. Tim Ball's meeting with the *Ottawa Citizen* editorial board in "Tim Ball: Wilful Disregard for the Truth," DeSmogBlog, August 30, 2006, www.desmogblog.com/tim-ball-wilful-disregard-for-the-truth.
5 High Park Group Registration, Corporations Canada, http://strategis.ic.gc.ca/cgi-bin/sc_mrksv/corpdir/dataOnline/corpns_re?company_select=4326741#directors.
6 James Hoggan, "NRSP Controlled by Lobbyists," DeSmogBlog, January 18, 2007, www.desmogblog.com/nrsp-controlled-by-energy-lobbyists.
7 Government of Canada, Lobbyists Registration Regulations Questions and Answers, http://ocl-cal.gc.ca/eic/site/lobbyist-lobbyiste1.nsf/eng/h_nx00279.html.
8 The Heartland Institute, Video Record, 15 minutes, 57 seconds, www.heartland.org/bin/media/newyork08/audio/Monday/harris.mp3.
9 See note 1.
10 Kevin Grandia, "Elections Canada drops Friends of Science Investigation," DeSmogBlog, September 23, 2008, www.desmogblog.com/elections-canada-drops-friends-of-science-investigation.

CHAPTER 6

1 Frank Luntz interviewed by John McCaslin on the C-SPAN book review show *After Words*, January 29, 2007, www.booktv.org/program.aspx?ProgramId=7872&SectionName=.
2 "Pipeline Damage Assessment from Hurricanes Katrina and Rita in the Gulf of Mexico," DNV technical report, January 22, 2007, www.mms.gov/tarprojects/581/44814183_MMS_Katrina_Rita_PL_Final%20Report%20Rev1.pdf.
3 Naomi Oreskes, "My Facts Are Better than Your Facts: Spreading Good News about Global Warming," in *How Do Facts Travel?* (Cambridge University Press, forthcoming Fall 2009).

CHAPTER 7

1 Letter from Steven F. Hayward and Kenneth Green of the American Enterprise Institute for Public Policy Research to Prof. Steve Schroeder of the Department of Atmospheric Sciences at Texas A&M University, dated July 5, 2006, posted on the DeSmogBlog, www.desmogblog.com/sites/beta.desmogblog.com/files/AEI.pdf.
2 Kevin Grandia, "Leaked Memo Claims that GM, Ford Financed Pro-CO2 Ad Campaign," DeSmogBlog, July 27, 2006, www.desmogblog.com/industry-memo-claims-controversial-ad-campaign-financed-by-gm-and-ford.
3 Quoted at www.sourcewatch.org/index.php?title=Heartland_Institute.
4 See http://store.junkscience.com/apocalypse-no-why-39global-warming39-isnot-a-global-cr.html.
5 Original document pdf, open letter from Heartland Institute Senior Fellow, Environmental Policy James M. Taylor, www.realclimate.org/docs/Heartland.pdf.

CHAPTER 8

1 H. Josef Hebert, Associated Press, "Jokers Add Fake Names to Warming Petition," *The Seattle Times*, May 1, 1998, http://community.seattletimes.nwsource.com/archive/?date=19980501&slug=2748308.
2 Christian Jensen, "Re: Fred Singer's Comment on Trenberth's Article," Memo to *Natural Science*, February 11, 1998, http://naturalscience.com/ns/letters/ns_let08.html.
3 James Hoggan, "Signatory Bails on Anti–Climate Science Petition," DeSmogBlog, April 18, 2006, www.desmogblog.com/signatory-bails-on-anti-climate-science-petition.
4 "Swift Boat Veterans for Truth," posted on the SourceWatch Web site, undated, www.sourcewatch.org/index.php?title=Swift_Boat_Veterans_for_Truth.
5 "Oil and Gas: Top Recipients," on the Web site OpenSecrets.org, undated, www.opensecrets.org/industries/recips.php?ind=E01&cycle=2002&recipdetail=S&mem=Y&sortorder=U.

CHAPTER 9

1 Andy Rowell, "Hard Rockers: The Views of the Green Lobby Should Be Challenged, according to a new alliance," *The Guardian*, July 11, 2001, www.guardian.co.uk/Archive/Article/0,4273,4218957,00.html.
2 Mike De Souza, "Canadian Schools Sent Brochures, DVDs from Climate Change Skeptics," *The Vancouver Sun*, May 4, 2008, www2.canada.com/vancouversun/news/story.html?id=528f1979-0c8e-4900-96d5-e586ac2fc435&k=55700.

CHAPTER 10

1 "Commitment to Development Index 2008: Denmark," on the Center for Global Development Web site, www.cgdev.org/section/initiatives/_active/cdi/_country/denmark.

CHAPTER 11

1 Ben Goldacre and David Adam, "Climate Scientist 'Duped to Deny Global Warming," *Guardian*, March 11, 2007, www.guardian.co.uk/media/2007/mar/11/broadcasting.science.

CHAPTER 12

1 "The Trashman Speweth," on the Center for Media and Democracy Web site PR Watch.org, undated, www.prwatch.org/prwissues/1999Q4/avery.html#milloy.

CHAPTER 13

1 H. Joseph Hebert, "Oil Money Favors Bush, Sees Gore as Too Contentious," *The Seattle Times*, July 3, 2000, http://community.seattletimes.nwsource.com/archive/?date=20000703&slug=4029919.
2 Philip Shabecoff, "Global Warming Has Begun, Expert Tells Senate," *New York Times*, June 24, 1988, http://query.nytimes.com/gst/fullpage.html?res=940DE7DF133AF937A15755C0A96E948260.
3 Environment Canada, *Canada's Greenhouse Gas Emissions: Understanding the Trends*, November 2008, www.ec.gc.ca/pdb/ghg/inventory_report/2008_trends/2008_trends_eng.pdf.
4 Ibid.

CHAPTER 15

1 Mike De Souza, "Billions of Litres of Tainted Oilsands Water Leaking: Report," Canada.com, December 9, 2008, www.canada.com/topics/news/national/story.html?id=1049842.
2 Jason Markusoff, "Alberta the Underdog in Oilsands PR Battle: Stelmach," *The Vancouver Sun*, April 30, 2008, www2.canada.com/vancouversun/news/story.html?id=a8205b8c-c669-4712-8524-e1aa24cb2163&k=86213.

INDEX